QUANTUM
LIFT

QUANTUM
LIFT

OUR POWER TO RESTORE INFINITE POTENTIAL

IN A QUANTUM REALITY

BY KRISTINA DE CORPO

DEDICATION

This book is dedicated to all of us

CONTENTS

ACKNOWLEDGMENTS

I would like to acknowledge my husband, Sam Gomez and my three children, Roman, Lola and Mars, who never stopped believing in me, as I chose to dive farther and farther into what seemed impossible.

I would not have leaped into the written expression of my ideas without the encouragement of Rudolph Bauer. His consistent encouragement to document my theoretical thinking was the catalyst for me to rationalize and publish my ideas. I am grateful for his determination.

I would also like to thank Ediho Lokanga. He wrote to me after I posted my first paper on academia.edu and though I had no formal training, Ediho saw a vision within my work that encouraged me to keep going.

Another person I would like to thank is Wram Accorsi. It was his message of encouragement to apply mathematics to my theoretical thinking that allowed me to take the greatest leap in understanding. He stated his encouragement with clarity and conviction, which opened my mind and my heart to the simplest mathematical expression of my ideas.

I also would like to acknowledge Gustaz Szigligetius. I have a deep fondness and appreciation for this source of support. It is when we are willing to become our most authentic self that we find the solutions are within us. I thank Gustavo deeply for our meaningful communication.

The last person I would like to thank is my dearest friend, Tracy Lynch. It is when we have a source of belief that is stronger than our doubt, that we find our way back to who we really are as whole resonance, in order to have the courage to become it. My deepest gratitude to Tracy, for never once doubting my ability and being a constant source of love and support.

However, my biggest acknowledgment is to our shared reality, for without all of us having this experience, I could not be having mine. I am grateful for the human race. I believe in all of us, as whole integers relaying through time and space, at a single fixed rate of speed breaking the barrier of sound repeating within itself, as any combined possibility of ourselves, infinitely repeating.

INTRODUCTION

I am a creative strategist. As a strategist, I look at what wants to become possible within an organization, and I strategize ways for that organization to experience successful outcomes, based on the information being experienced.

My passion is systemic social change. As a strategist, I focus on why we are experiencing chronic social problems, in order to better understand how to systemically solve them. I entered into exploring quantum reality, because I saw an increasing polarity in our world. And I saw a gap in physics or an understanding of how reality works. It made me question, if we are able to explain one, would it explain the other?

When I took a deeper look into general and special relativity, I did not feel my experience of reality reflected. In terms of how I see and feel everything as connected from a place of oneness, or out of one. And the quantum world has been described over and over as weird in how it works. But in all my time in nature, I have never thought nature to be weird. To me, nature is filled with relativity, purpose and meaning.

Nature is always complete as a system, in the sense that nature relies on synergy to thrive as a system. When nature does not thrive, there is always a missing condition that allows nature's own ability to restore balance, as the whole repeating to become greater than the sum of its parts. Meaning, the equilibrium of the whole system has been destroyed, over retained.

Relativity and quantum theory do not work as a whole system. Meaning, they each are successful on their own, but do not work synergistically together. Saying it a different way, when these two theories interrogate through equations, the information they express does not relate for a greater picture to be understood. Equations allow what was once previously described, to become undeniably clear, through the repeat application of a fixed metric.

1

Saying it yet another way, if relativity and quantum theory work synergistically, but do not isolate compound integers, something seems to be missing, in order to provide a working model of *how* the Universe isolates oscillating frequency, as dimensional particle reality. This led me to consider if there is a fundamental way of restoring reality, which rasterizes or converts 100% of lost wavelength, through these two experimentally proven theoretical approaches to reality. Which allowed me to consider, are we rasterizing or converting 5% of lost wavelength, which would exhibit force within unequal amounts of still motion, as normal matter?

I have come to the conclusion *"we"* are missing, in terms of how we as human beings exhibit properties of oscillatory motion, known isometrically as **x** and **y** coordinates, in order to rasterize or convert *perceived* wavelength, into our dimensional particle reality. In other words, do **x** and **y,** exhibit mutually interchangeable properties, to coordinate horizontal and vertical reality? If one continuous line of still motion does not exhibit latent properties, as properties concealed but not yet manifested, straight motion is not simultaneously repeated, one vertical coordinate does not *rotate,* in order to restore a horizontal axis, as exhibited rationalized potential, over lost wavelength.
It is now we must consider if a fixed or pre-determined rate of oscillatory motion, which would allow human beings, as a rate of change as derivatives, to ascend at a single fixed rate of isolated potential, as a pattern of fixed percentages. To do this, we must separate the raw from the mean by allowing only one *ascending* amplitude to **synchronize** mutually inclusive properties, as repeat interchangeable properties. Meaning, one ascending charge, as a single fixed rate of oscillatory motion, would ascend at the apex or rest point of oscillatory motion, which would oscillate a substantially larger picture.

In other words, I believe we are only rasterizing or converting "part" of the missing wavelength or frequency, as dimensional particle reality, which distorts the picture or mitigates the charge,

2

as lost or *descending* wavelength cannot simultaneously repeat, if a single fixed rate of oscillatory motion is not repeated.

However, within mutually interchangeable properties as isolated potential, as the simultaneous conversion of both **x** and **y** coordinates, within a single fixed rate of percentages, it is possible to isolate a single fixed rate of charge, as a single solution auxiliary field, as repeat pattern.

Saying it a different way, to *isolate* an oscillating quantum reality requires a pre-determined fixed set of rationalized percentages, as a repeat pattern. Meaning, you cannot simply take a fraction of a number and simultaneously repeat it, as a single solution auxiliary network. Put simply, you must "isolate" a raw rate of pre-determined motion, as a single solution with zero opposing amplitude. In order to do this, we must separate the raw from the mean, or the exhibiting properties from the latent properties.

Saying yet another way, we must *separate* what exists now, as fixed properties, as a simultaneous repeat of interchangeable properties, out of one straight line of motion with no stopgap or *disruption* of oscillatory motion. Which only occurs at the rest point or apex of two mutually inclusive properties, as an ascending or single solution auxiliary field, as repeat pattern.

In other words, you cannot *derive* one mutually interchangeable isotope, as a compound integer, as a single fixed rate of fixed percentages, if latent properties do not exhibit equal amounts of force, as a single fixed rate of balanced pressure. Put simply, light **refracts** within *equal* measure of volume of sound repeating, as a single isotope. We then rasterize or convert, dimensional particle reality, as mass, at a single fixed rate of isolated potential, with no destruction of charge, repeatedly.

Relativity must be measurable, as an **isolated** repeat pattern, in order to equate one continuous straight line of oscillatory motion, as a single solution auxiliary field, as repeat pattern, rasterizing or converting dimensional particle, as mass. In other

3

words, a single fixed rate of isolated potential, must *release* a positively charged positron to replace compound integers at the gap of utility, when applying a metric-based equation. Otherwise, we cannot calculate the *loss* of electrons destroying lost or descending wave frequency, which *disturbs* still motion at the rest point or the apex of oscillatory motion.

Becoming a **master** of quantum reality, means to *unilaterally* distribute a single fixed rate of isolated charge, as an ascending auxiliary feed, throughout the *whole* repeating oscillation, in order to isolate a quantum possibility resonating with you. To do this, you must restore "conciliatory rotation" restoring your lost wavelength, as mutually interchangeable properties.

In other words, oscillatory motion, at the rest point or the apex of the wave redistributes charge, *magnetically.* Which means motion, as one continuous straight line, is only redistributed at the rest point, if a repeat pattern subdivides itself, *isometrically.* Meaning, one continuous straight line of motion exhibits itself isometrically, as future input for all preceding operations, if the motion is not disturbed at the rest point or apex of oscillatory motion. For the purposes of this book, one continuous straight line of motion relaying within mutually interchangeable properties is called **conciliatory rotation.**

Throughout this book, I will guide you through this new theoretical understanding of how we as human beings experience a quantum wave pattern, as one continuous straight line of motion, with no stopgap, or disruption of motion. I will share what I believe is the factor that either limits or allows, a single fixed rate of conciliatory rotation, or **A** to propagate **B,** as one continuous rate of oscillatory motion, rationalizing a single fixed rate of isolated potential, without mitigating charge or "shortening" descending lost wavelength. Which is the fundamental reason why I believe we only view 5% of isolated oscillatory motion, as a single fixed rate of charge, as normal matter reflecting less than a single fixed rate of charge.

Oscillating frequency must be pre-determined or fixed, as a rate or amount of *one* full rotation, at the rest point or the apex of oscillatory motion, in order to exhibit latent properties, rasterizing or converting, a single fixed rate of charge, as isolated potential. Which only then **repeatedly** allows for a single fixed rate of oscillatory motion, at the rest point or apex of *conciliatory rotation.* But only within mutually interchangeable properties exhibiting equal amounts of oscillatory motion. This means the speed of light displaces 100% of the rest point of oscillatory motion, without decreasing the apex, or volume of space, which then does not divide one continuous rate of oscillatory motion, as a single fixed rate of charge.

I will be presenting the **one electron theory** in the second chapter. And I will slowly build up to the understanding of the **theory of everything based on balance and harmony,** in order to theoretically be applied to our experience of a quantum field, as infinite potential, by the end of this book.

This book is not a compendium of scientific research, to the extent that it is a field journal. It has been intuitively designed to isolate a rational approach to interrogating two very fundamentally different, but theoretically proven, approaches to constructing reality *through* isolating "one" as a single repeating metric. In other words, the interchangeable properties of 100% of one *continuous* or straight line of motion, exhibit mutually inclusive properties, but only when the rest point at the apex of conciliatory rotation, is not disturbed or *disrupted.*

The rationale for one continuous line of motion is clear when we determine *how* the quantum Universe propagates a single fixed rate of repeating charge. Meaning, as a single fixed rate of oscillatory motion, as a fixed or pre-determined amount of the percentage of conciliatory rotation, raw potential is "fundamentally" restored. Which means mass must exhibit the latent properties, as mutually interchangeable properties, at a single fixed rate of speed, as charge repeating, at the rest point or apex of conciliatory rotation. This would allow the speed of light to oscillate as mutually interchangeable properties, as an *ascending* quantum wave pattern, with zero electrical potential difference.

Put simply, charge must be united as *one* non-exclusive property of oscillatory motion, in order to be latent within any mutually interchangeable properties. In order to recontextualize what it means to be human, we must consider being all the fundamental properties that restore a single ascending amplitude, restoring a fixed rate of percentages as lost or descending wavelength, with zero electrical difference. What we must now reconsider is what causes the loss of oscillatory motion, at the rest point or the apex of conciliatory rotation?

Distorted wavelength must ascend as a *positive* charge, as rations of one single amplitude or fixed rate of potential, but only if lost wavelength retains a single isotope, as a pre-determined amount of conciliatory rotation, in order to achieve infinite mass, as isolated potential. Meaning, *through* its magnetic property, **helium** *restores* lost or *descending* sound waves, as a magnetic shield restoring mutually exclusive properties. We must consider the mutually interchangeable properties of **sound,** as a single fixed rate of charge, at the rest point of conciliatory rotation, as a pre-determined amount of 8 fixed rotations, which simultaneously *restores* **oxygen.**

In other words, a single isotope is retained, over transferred, to isolate balanced pressure, as oscillatory motion releases a single fixed rate of charge, as a fixed or pre-determined amount of conciliatory rotation. Which synchronizes the restoral of mutually interchangeable properties.

To do this, we must now invert the Pythagorean theorem to *retain* whole ratios, at the rest point or apex of conciliatory rotation. Which allows speed to accrue at right angles or fixed proportions of itself exhibited only at the **y** axis point, which fundamentally restores itself as **x**. Because in order to convert rations of one, as a single repeating metric, into whole percentages of itself repeating, as a single solution auxiliary network, a single isotope must retain *charge, as a positively charged positron.*

Which would mean it is possible to separate the latent properties from the exhibited properties. This means slope cannot be pre-determined, unless a rationalized whole integer fundamentally restores itself, within a *single* isotope. In other words, one continuous straight line of motion is positively charged, at the rest point or the apex, within mutually interchangeable properties.

To truly isolate infinite potential, in a quantum particle reality in three dimensions, we must isolate a single fixed rate of potential, as a pre-determined amount of conciliatory rotation, to isolate raw potential. Infinite potential is not endless. It exists as mutually interchangeable properties, until 100% of *one*, as a single fixed rate of raw potential, can no longer derive itself, within a pattern of fixed percentages. In other words, are we exhibiting the factor which would allow us to ascend as one continuous straight line of undisturbed motion, at the rest point or apex of conciliatory rotation, which would allow the subatomic world to become rationalized as whole integers? Or are we disturbing the motion?

Any current astrophysics journal repeatedly states, we currently only visually experience 5% of the quantum Universe as "normal matter". But what we have not considered, is 5% *less* than 100% of mutually interchangeable properties, does not simultaneously equal itself, infinitely restoring itself, *isometrically.*

I believe we are only rasterizing or converting 5% of synchronized reality. But in order to exhibit latent properties, the restoral rate of oxygen, as a single fixed rate of restorative properties, must not **exceed** 100% of one continuously charged isotope, *or the quantum wave pattern will retract.*

What I am suggesting is that the subatomic rotation of oxygen, as a single fixed or pre-determined amount or rate of conciliatory rotation, does not exist as a mutually interchangeable property, within a single lost isotope that does not simultaneously replace itself, as a positively charged positron. In other words, the release of an electron occurs *twice,* **with no stopgap** or *loss* of oscillatory motion, but only at the rest point or apex of conciliatory rotation, does it simultaneously restore itself. However, one isotope must simultaneously *retain* or **magnetize** one continuous straight line of oscillatory motion, at the rest point or apex of conciliatory rotation.

Meaning, space and time are not relative, in so much as their physical properties are *isometric*, **within the geometry of one simultaneously repeating a single isotope, as a positively charged positron.** Only then, within a rationalized ascending harmonized standing wave pattern, will a single fixed rate of charge, as a pre-determined amount of 8 subatomic rotations as oxygen, not disturb the rest point or apex of conciliatory rotation. Which simultaneously repeats and mutually *extends* the boundary of time and space.

In a simple causality ratio, a single fixed rate of pre-determined amounts of conciliatory rotation would be modeled as **A + B** are mutually interchangeable, as *one* non-inclusive property, or single fixed rate of charge, isometrically repeating itself. Which restores oxygen subdividing itself 8 times, at the rest point or apex of conciliatory rotation, as a single isotope. However, if **8** is not *subdivided* within a single fixed rate of percentages, **A** destroys **B**. Meaning, reality is absolute, but only in the sense that a single fixed rate of charge must simultaneously *repeat* itself, as it mutually extends the boundary of time and space, or **A** will destroy **B**.

To put it simply, dimensional particle reality as mass is an isolated system, as one continuous straight line of motion exhibits itself as a single solution auxiliary field, as a repeat pattern.

To recontextualize this understanding more, if one single fixed rate of charge *simultaneously* repeats a 100% of one single isotope, exhibiting force as one continuous line of motion, with no stopgap or disruption of oscillatory motion, at the rest point or apex of conciliatory rotation, a third coordinate **z** is pre-determined.

However, in order to fundamentally *restore* lost or descending wavelength, within mutually non-exclusive properties *extending the boundary of time and space*, one single fixed rate of isolated charge must propagate itself, as a single solution auxiliary field, as a repeat pattern. Which pre-determines a single fixed rate of conciliatory rotation, at the apex or rest point of oscillatory motion, in order to not disturb a positively charged *neuron.* Because we *are* the **processors** of our own reality.

Meaning, the pattern must *ascend* in order to release and simultaneously retain a single isotope, at the rest point or apex of conciliatory rotation. This can only happen if one positively charged positron is released and simultaneously retained, to occur **twice.** Meaning, if a **right angle** *precedes* distorted wavelength, the **x axis** will *retain* perpendicular drag, which allows 100% of thrust, as the restorative properties of the **y axis,** but only in a single solution auxiliary field, as repeat pattern.

One positron must exhibit an isolated or a single fixed rate of repeating charge, at the rest point or apex of conciliatory rotation, for oxygen to subdivide itself 8 times, as mutually interchangeable properties of one single *positively* **charged** positron.

It is the observation of the physical nature of our reality that informs our theoretical thinking. I am exhibiting a whole new way of comprehending how we fundamentally restore dimensional particle reality, as mass. Which may, or may not, exist in this collective version of multidimensional reality. In other words, we are not confined to the chemical makeup of the physical properties of our bodies, as mass.

This book is for those seeking a new way to understand *how* we as human beings experience a quantum reality. Our quantum world allows for infinite possibilities. It is not just observable reality. Reading this book takes a commitment to reconsider our approach to making sense of the physical properties we observe and isolate, as the *three* fundamental exchanges that comprise our dimensional particle reality, as length, width and height, as mass.

To facilitate the comprehension of this understanding, the start of this book is complex, in order to recontextualize a greater understanding of our human experience of quantum reality. There is also a glossary to help guide you through assimilating the ideas presented in this book.

It is possible for relativity and quantum theory to mathematically or *magnetically* interrogate. But only if 100% of one restores a single fixed rate of charge, simultaneously, as a pre-determined amount of oxygen subdividing itself 8 times, as a repeat pattern synchronizing both the simultaneous release and repeat of a positively charged positron. Which must *isolate* a single fixed rate of charge, as a single solution auxiliary field, as repeat pattern *restoring* a **tri-auxiliary network.**

There is a cosmological constant we must understand, in order to access our quantum power in a dimensional particle reality that *instantly* restores lost potential, *faster* than the speed of light. Meaning, one positively charged integer both ascends and **simultaneously** *restores*100% of one single isotope, as a single amplitude exhibiting latent properties for future input of all preceding operations. However, this does not mean a positively charged positron remains charged, unless 50% of the apex of conciliatory rotations has decreased.

One positively charged positron is <u>not</u> *complete*, if a single fixed rate of charge, as a pre-determined amount of conciliatory rotation, is *not* fundamentally restored, as any mutually interchangeable property. Which means a **positive feedback loop,** for future input of all preceding "differentials" or differing amounts, is synchronizing a single fixed rate of charge, as a pattern of fixed of percentages.

10

Reality is a restorative process, within the only law which allows any repeat matrix or wave pattern to ascend, within *one* mutually reinforced **single** amplitude. In other words, we will be looking at the conservation of energy, as a principle, to reconsider the law of fixed percentages, as a fixed rate of *proportions.* This is an intuitive approach to the art of storytelling what particle reality would be, if we were all forced into our own unique resonance, as natural harmonic frequencies mutually or independently restoring our lost wavelength. As a single fixed or pre-determined rate of conciliatory rotation, magnetically restoring oxygen, as a single "rotating" isotope, or positively charged positron.

This book is designed to promote rational thinking by inviting discourse regarding how we as human beings rasterize or convert an oscillating reality. Almost all or 100% of this book is common knowledge, which I have adapted into a single solution auxiliary field, repeating an isometric pattern restoring fundamental properties. Therefore, I have chosen to bookmark any source needed to be noted, as a bibliography in the back of the book.

This is a position I invite you to explore, in order to be persuaded it is possible to isolate preceding wave pattern, as one continuous straight line of motion simultaneously restores itself, as mutually interchangeable properties. As a single solution auxiliary field *retaining* oscillatory motion, at the rest point or apex of conciliatory rotation. Which does not disturb the rest point of one continuous straight line of motion, unless the motion has been disturbed, which decreases the apex and *releases* ionic pressure.

We are infinite, but only within combined possibilities of 100% of one restoring itself, isometrically, or within equal amounts of volume of space. Meaning, until we reach the **point** of *dividing* mutually interchangeable properties, we are any combined possibility, infinitely expressed as whole integers.

PART ONE

YOU WITHIN ISOLATED POTENTIAL

1

YOU AS A DERIVATIVE OR RATE OF CHANGE

THE BUILDING BLOCKS OF QUANTUM REALITY

There is no limit to this quantum reality and yet our current reality does not feel as if this is true. What I am suggesting is the idea that you are a derivative, as a whole pre-determined wave cycle, which oscillates as infinite possibilities.

Meaning, you have the fundamental restorative power to exhibit pre-determined properties, as dimensional particle reality, as mass. Which means at the cellular level, we are a restorative property of any combined possibility. But we also have the *destructive* power, to **cancel** any combined possibility, as probability.

In a quantum reality, one continuous straight line of motion must subdivide itself, in order to mutually *extend* wavelength, as mutually interchangeable properties, within the boundary of space and time, over exclude properties. But at the same time, light must simultaneously repeat itself, within mutually interchangeable properties, in order to exhibit force, within synchronized pressure, to oscillate **x** and **y** as one simultaneous coordinate.

In other words, unless we repeat a pre-determined oscillation, there is not an infinite amount of possibilities for us to fundamentally isolate our dimensional particle reality, within equal amounts of oscillatory motion. An auxiliary field is one whose equations of motion admit a single solution, with zero opposing amplitude or electrical potential difference.

Meaning, one continuous straight line of motion does not stop, under a repeating isolated metric, which allows the wave to distort or shorten, as it simultaneously repeats one full rotation. Saying it a different way, equal amounts of oscillatory motion, as a single fixed rate of charge, can either be a negatively charged positron, or a *positively* charged neutrino, unless 100% of one is retained as a positive feedback loop for future input of all preceding differentials, as their own operators.

SYNCHRONIZING INFINITE REALITY

The big question then becomes, how do we know if we are oscillating within isolated potential, as a single fixed rate of charge, simultaneously restoring 100% of itself, as mutually interchangeable properties, over mitigating it? Saying it another way, if a wave can isolate particle reality, out of oscillating frequency, as a single repeating charge, do we understand the condition limiting us from isolating equal amounts of ionic pressure? Both individually and as a collective?

In other words, are we looking at isolating equilibrium as a factor? Which would propagate a single fixed rate of charge, as a repeat pattern establishing a single solution auxiliary field. Meaning, can we isolate the factor which repeats the pattern, as a single fixed rate or amount of conciliatory rotation, to retain balanced pressure? Put differently, can we restore isolated potential, as mutually inclusive properties, in such a way that lost potential becomes *absolute* or **fixed,** as the completion of oscillatory motion, as one continuous straight line of motion, over probable? Which mitigates the charge, as lost or *disturbed* motion, as descended wavelength.

We must understand, there is a preceding factor that *disallows* the visual rasterization of isolated potential, as one whole complete charge or rotation of oscillatory motion fundamentally restoring our lost wavelength, as dimensional particle reality, as mass.

Energy must be released and restored in equal amounts, to retain balanced pressure in a single solution auxiliary field. It is critically important that we realize there is a fundamental way of seeing reality, when force is exhibiting equal amounts of pressure. Any combined possibility works, which we can then see within the complete restoral of 100% of lost wavelength. Over 5% as the outcome of mitigating or disturbing one continuous straight line of motion, without *restoring* 100% of it.

The law of conservation of energy validates that any amount of oscillatory motion **must** *amplify* 100% of any distorted wavelength, ascending as a **repeating** or positive feedback loop, unless the motion is exhibited on by an outside force. Which means *two open* systems, restore **one** harmonic frequency as pattern. Which must allow natural harmonic or resonant frequencies reoccurring within the amplitude of **one,** as a single repeating harmonized standing wave pattern, as one whole complete oscillation or fixed *rotation.*

Which must mean 100% of one repeatedly isolates resonate frequency rasterizing or converting itself into dimensional particle reality. **Resonate frequency** is when balanced pressure is applied in harmonic proportion to the natural frequency or repeating rhythm of a system, as a simultaneous *pattern* restoring lost wavelength.

In a quantum reality, *observation* is what causes oscillatory motion to continue, as one full "complete" rotation, which *constructs* dimensional particle reality, out of the geometry of one, restoring itself, *proportionately.* As a single fixed rate of repeating charge, as one continuous line of oscillatory motion, at the rest point or apex of *conciliatory rotation.* Which now becomes the full repeat oscillation, synchronizing a fixed amount of conciliatory rotation, to fundamentally restore 100% of itself, as oscillatory motion, rasterizing or converting dimensional particle reality, as mass.

To make the distinction, *mass* is the information we experience as real. In other words, mass is our physical or *dimensional* experience of what is, right now, in the present moment, as particle reality. Which is the outcome of force, as one continuous straight line of motion exhibits itself as mutually interchangeable properties, by disturbing the motion at the rest point of oscillatory motion, without disrupting the motion or decreasing the apex of conciliatory rotation, as mutually exclusive properties.

Matter or fundamental mass is the *potentiality* of one continuous straight line of oscillatory motion **simultaneously** restoring 100% of one single isotope. Which fundamentally restores itself as mutually interchangeable properties, by repeating a single isotope if motion is continuous, or releasing a single isotope, if motion is disrupted. This is seeing right through the paradox of physics.

Meaning, **mass** as dimensional particle reality can rasterize or convert any state of itself, out of fundamental mass simultaneously restoring itself, as a single fixed rate of charge, as a single solution auxiliary field, as repeat pattern. Until you observe it. Then your oscillation must either retain equal amounts of oscillatory motion or do the mechanical work to retain the geometry of one, repeating as a single isotope.

Simply put, if you do not retain oscillatory motion as one continuous straight line of motion, exhibiting equal amounts of pressure within repeating volume of space, you *resist* a single fixed rate of charge at the cellular level. Which you then are unable to synchronize ascending harmonized standing wave pattern, as a single solution auxiliary field, as a positive feedback loop.

Which means you cannot isolate dimensional particle reality as mass, out of less than 100% of your resonant frequency, exhibiting force within balanced pressure, to fundamentally restore itself, as any combined possibility, within mutually interchangeable properties. This is because your resonant frequency descended into *multiple* vertices, as an **incomplete** conciliatory rotation, at the rest point or apex of oscillatory motion.

From the beginning of the Universe, to the end of the Universe, resonate frequency oscillates as a single fixed rate of charge, exhibiting itself as one continuous straight line of motion, as a positive feedback loop isolating or restoring itself synergistically, in order to oscillate as anything different.

A feedback loop is a system retaining 100% of mutually interchangeable properties, exhibited as 100% of one single fixed rate of a pre-determined amount of conciliatory rotation, as one continuous straight line of motion, as a future operation for all preceding differentials. In a quantum reality, it is whether the complete oscillation restores lost potential, as an ascending wave pattern, within a single ascending amplitude, at a single fixed rate of charge. Which completes a positively charged positron as a feedback loop restoring a distorted horizontal axis, and wavelength is restored through the process of ionization, over cancelled.

This is why in the well-known double-slit experiment[1], when there is only **one** slit, oscillation forms a **single** band of light. But when there are **two** slits, oscillation forms an *interconnected* wave pattern. Within one slit, motion is disturbed. Within two slits, anything can visually reoccur, out of isolated potential within one continuous line of motion, as the whole repeat oscillation.

ISOLATING POTENTIAL AS DIMENSIONAL PARTICLE REALITY

To define conciliatory rotation in simple terms, it is all the pre-existing co-factors of one, in any amount, that do not disturb one continuous line of motion, at the rest point or apex of oscillatory motion. It is the law of fixed percentages that disallows this rate of fixed proportions. To arrive at the greater understanding, we must take 100% of one and subdivide it equally three times, which rationalizes a third infinitely, with only .01% remaining. 100% of one, is any mutually interchangeable property, as any whole number ratio.

The quantum wave must ascend within a single isotope that mutually extends the boundary of space and time, as mutually interchangeable properties, and at the same time, coincide with itself within any measurable difference, as the volume of the geometry of one, repeating a positively charged positron. Which allows does not exclude *interchangeable* properties. We must consider the three coordinates of **x, y** *and* **z** coinciding as mutually interchangeable properties, as a single isotope, to simultaneously *project* any measurable difference.

Let us reconsider *how* a harmonized standing wave pattern repeats oscillatory motion. Any whole number can be rationalized and ascend within mutually interchangeable properties. It is when properties are mutually exclusive, that the pattern breaks up and whole integers isolate proportionately *less* of a single fixed rate of isolated percentages. In other words, to mutually ascend, one must infinitely repeat itself, within a 3 to 1 ratio, in order to exhibit pressure, within an equal amount of force, to *retain* a single isotope.

I believe we must isolate, an oscillating quantum reality, as one continuous straight line of motion, with no stopgap. In order to simultaneously repeat mutually interchangeable properties, which also simultaneously *extend* missing wavelength, as mutually exclusive properties, as the conversion rate to fundamentally restore dimensional particle reality, as mass.

Meaning, at the same time and moment in space, the properties of the **x axis** must simultaneously repeat itself, within mutually non-exclusive properties, in order to exhibit interchangeable properties synchronizing **x** and **y** *coordinates*. At a single fixed rate of oscillating potential exhibiting *both* the **x** and **y** axis, as a *single* **point**, of mutually inclusive properties. Which coordinates a third "missing" or pre-determined coordinate **z**, as a visual rasterization or conversion of a fundamentally restored dimensional particle reality, within the geometry of one, or 100% of one, as a simultaneous repeat of itself, *proportionately*.

In other words, unless we *magnetize* a pre-determined or single fixed rate of oscillatory motion, at the rest point or apex of conciliatory motion, there is not an infinite amount of possibilities for us to isolate, as our dimensional particle reality, as a single repeating isotope exhibits one continuous line of oscillatory motion.

An auxiliary field is one whose equations of motion admit a single solution, with zero opposing amplitude or electrical potential difference. Meaning, a single line of motion does not stop, under a repeating isolated metric, as it simultaneously extends itself. Which allows one continuous line of motion to distort or shorten, as it simultaneously repeats one full or single fixed rate of charge, as a pre-determined or fixed amount of rotation, amplifying 100% of its subatomic properties. Receiving isolated charge does not disturb the motion, at the rest point of conciliatory rotation, if both properties are latent within one pre-determined single isotope.

The notion of time and space then becomes pre-determined, as the fixed *non-dimensional* properties of 3-D particle reality, at the rest point or apex of conciliatory rotation, simultaneously restore a "dimensional" particle reality, as mass. But only within a **pre-determined** or *fixed* amount of conciliatory rotation, will pressure **equally** distribute a single fixed rate of charge, by simultaneously retaining and isolating a single isotope, as mutually interchangeable properties.

Meaning, ascending wave pattern **will** *retain* a fixed or pre-determined amount of one continuous straight line of motion, by not *excluding* mutually interchangeable properties. In order to exhibit a fixed rate of proportions, which magnetize or simultaneously retain a single isotope, as any rationalized compound integer, simultaneously repeating itself, as an ascending mutually *interchangeable* property.

Saying it a different way, one ascending wave pattern cannot repeat a pre-determined rotation, as an auxiliary field which equates all motion, as a **single** solution. Unless a single or continuous straight line of motion is not disrupted at the rest point or apex of conciliatory rotation.

Mutually exclusive properties include the non-localized "entangled" particle but must retain one continuous straight line of motion, with no stopgap, or disruption of motion, in order to simultaneously restore it. If mutually interchangeable properties do not retain a single fixed rate of charge exhibiting pressure within equal amounts of volume of space, at the rest point or apex of conciliatory rotation, the wave retracts, as the pattern cannot restore itself, within less of itself, proportionately.

Put more clearly, **oxygen** must *retain* **8 subatomic rotations,** at the rest point or apex of oscillatory motion, to release a single isotope which is mutually interchangeable, as a property of both helium and hydrogen, as mutually interchangeable properties.

Put simply, the chemical foundation for an oscillating quantum reality that isolates a single fixed rate of percentages of itself, repeatedly restores an optical awareness of a single solution auxiliary field, as repeat pattern.

In other words, the mutually interchangeable properties of **one** *single* isotope, rasterizing or converting itself, within mutually interchangeable properties, extends the boundary of time and space, as non-exclusive properties. Which means a single fixed rate of charge, as one continuous straight line of motion exhibits *equal* amounts of oscillatory motion, without disturbing the rest point or apex of conciliatory rotation.

We must reconsider the law of motion to include a law of fixed percentages that states, any line of motion cannot supersede any mutually co-existing or interchangeable property, without repeating the same rate of oscillatory motion.

Meaning, at the fundamental level, time and space *pre-exist,* within infinite fixed percentages of a single coordinate, simultaneously coordinating a calculable difference, within fixed proportions of itself, as mutually interchangeable properties. Unless one continuous straight line of motion, does not equally disturb force, as pressure, *within* mutually interchangeable compound integers, displacing 100% of one, simultaneously.

ISOLATING WHOLE RESONANT FREQUENCY

In order to consider *why* a mutually extended boundary destroys isolated pressure exhibiting interchangeable properties, as a single fixed rate of charge, we must understand how it repeats itself, as any whole or compound integer replacing a single isolated isotope, as isolated pressure.

How do you *perceive* this reality? Do you feel "one" with everyone and everything and include everyone and see your reality as mutually interchangeable properties? Meaning, within your visual experience of 5% as normal matter, as mutually exclusive properties, do you feel *limited* or **unlimited** as 100% of isolated potential, as a single fixed rate of charge, mutually extending the boundary of time and space as any combined possibility?

If not, this translates into **destructive wave interference,** which cancels balanced pressure out, by exhibiting force within a system, fundamentally restoring an isolated metric, as a single solution auxiliary field, as repeat pattern. Which then cannot retain a single "mutually" reinforced amplitude, as mutually interchangeable properties. Considering at the subatomic level, an ascending wave cannot exhibit a single fixed rate of charge, as an auxiliary field restoring a repeat *ascending* or harmonized standing wave pattern, unless a pattern restores itself *isometrically*, as mutually interchangeable properties.

To contextualize this understanding further, oxygen must exhibit 8 subatomic rotations, to exhibit 100% of thrust, within balanced pressure, or helium cannot derive a single fixed rate of oscillating potential, as a single isotope.

Meaning, you cannot isolate the restoral of fundamental mass, within dimensional particle reality, as mass, unless oxygen is retained in equal amounts. If you "cancel" lost wavelength out, it *disrupts* conciliatory rotation, at the apex which means mutually interchangeable properties descend within multiple vertices.

In other words, you *cannot* restore or **isolate** mutually interchangeable properties, as a single fixed rate of one continuous straight line of still motion, without disturbing the rest point of oscillatory motion.

Our perception of particle reality, as whole integers replacing a single fixed rate or pre-determined amount of a single "ascending" charge, at the rest point of oscillatory motion, is what causes a repeat harmonized standing wave pattern to ascend, within 100% of itself, by exhibiting mutually interchangeable properties. It is critical for us to grasp within our own awareness, and our own experience of it.

We are one continuous straight line of still motion, as our own unique resonant frequency, as **one** *ascending* harmonized standing wave pattern, deriving a single fixed rate of isolated potential. As whole integers repeating a pattern of whole number ratios, which must allow one continuous line straight line of motion, to repeat or ascend, within no stopgap or disruption of the apex or rest point of oscillatory motion.

This is our quantum power fundamentally realized. We can isolate and restore any solid, liquid, gas or plasma state within mutually interchangeable properties, as a mutually extended boundary of time and space, that does not exclude the non-localized particle. We are one quantum wave pattern isolating and restoring itself, as the mutually interchangeable properties of the speed of light traveling equidistant within a pre-determined or fixed rate of sound. Which means, we can synchronize the restoral of a single isotope simultaneously synchronizing 100% of the speed of light, pre-determining particle reality.

However, if one continuous straight line of motion is disturbed at the rest point of conciliatory rotation, oxygen does not rotate or complete 8 subatomic rotations, to subdivide itself as equal amounts of pressure, within repeat volume of space. Which then cannot replace the mutually interchangeable properties of sound exhibiting unequal amounts of force, which retracts or reflects the speed of light.

We all know hydrogen, in equal amounts, fundamentally restores the properties of oxygen, as water, or H_2O. It is the same for helium or light or sound, as mutually interchangeable properties. A single isotope must exhibit itself *once,* as one mutually interchangeable isotope which simultaneously restores lost wavelength, as the replacement of still motion, as 100% of itself, as it simultaneously disturbs the rest point of oscillatory motion. This is how we construct dimensional particle reality, without distorting lost wavelength, within rations of fixed properties or mutually interchangeable *ascending* frequencies.

You must avoid causing electrical potential difference or gap in volts, by *amplifying or continuously replacing* 100% of oscillatory motion at the rest point or apex, as one continuous straight line of motion, in order to exhibit equal amounts of pre-determined conciliatory rotation, as oxygen.

Only by simultaneously repeating mutually interchangeable properties *through* constructive wave interference, will oxygen simultaneously release and restore a positively charged positron. As an isolated system restoring lost potential, as equal amounts of mutually "inclusive" properties, to fundamentally restore resonant frequency.

THE FIRST LAW OF THERMODYNAMICS DOES NOT APPLY

So how do equal amounts of oscillatory motion apply to our experience of oscillating particle reality, as one straight continuous line of motion, with no stopgap or disruption of conciliatory rotation?

What it means is that if we take action as a derivative, our rate of change, as a single fixed rate of charge, either *ascends* or **descends,** as sequential order, which either exhibits a simultaneous **reaction,** as the restorative property, or the charge is mitigated by our choice within mutually exclusive properties.

In other words, one single isotope exhibits equal amounts of pressure, as the instant reaction or entangled particle at a *distance,* **locally.** Meaning, if the rest point of oscillatory motion is disturbed, one continuous line of motion is disrupted at the rest point of conciliatory rotation. Which means, heat is no longer being isolated.

Simply put, isolated potential, as a single fixed rate of charge, as a single solution auxiliary field must *retain* a single positively charged positron, as it simultaneously replaces a negatively charged neuron, as synchronized potential exhibiting pressure within equal amounts of isolated pressure. Meaning, if the fundamental property of oxygen is no longer exhibiting equal amounts of oscillatory motion, the geometry of one is disrupted, as one continuous straight line of motion. Which means, a pre-determined or fixed amount of conciliatory rotation does not release a single isotope, as it simultaneously charges a positively charged positron.

Any compound integer is replaced at once, within an ascending repeat harmonized standing wave pattern, as one whole simultaneous oscillation, exhibiting equal amounts of force, within *balanced* pressure. In other words, the *complete* rotation of isolated oscillatory motion in equal amounts, as a single fixed rate of charge, releases a pre-determined amount or fixed rate of oxygen, as force exhibited as 8 subatomic rotations, which must simultaneously release and retain a single isotope.

Saying it differently, mutually interchangeable properties only simultaneously repeat as 100% of isolated pressure, as a single fixed rate of speed, if oxygen is restored at its subatomic level of **8**. Meaning, isolated potential simultaneously repeats lost wavelength, within any whole integer, as one continuous straight line of motion, which does not decrease the apex of conciliatory rotation, at the rest point of oscillatory motion. Simply put, if isolated pressure restores oxygen, it will not disturb the **still point** of conciliatory rotation.

Simultaneity is the *simultaneous* repeat between any **two** ascending frequencies exhibiting *equal* amounts of oscillatory motion, which either mutually extends or *exhibits* latent properties, which rationalizes any whole integer. Meaning, all action occurs at once, as a simultaneous reaction, within equal amounts of force within balanced pressure. However, if force divides equal pressure, oscillatory motion is unstable, as descending lost wavelength within multiple vertices. Which means, *simultaneity* is not possible.

If one single ascending amplitude, as one continuous still line of motion, ascends within mutually interchangeable properties, force exhibits balanced pressure, within any *whole* integer, as a single repeating isotope, as a single fixed rate of charge. Which means oxygen subdivides mutually interchangeable properties, as isolated or fixed amounts of sill motion, replacing the rest point or apex of conciliatory rotation. In other words, subatomic energy is retained within the system, as the instant or simultaneous entangled particle at a distance, is restored.

Which restores any compound integer, as mutually interchangeable properties of any subatomic property exhibiting equal amounts of force, within balanced or isolated pressure. In other words, *oxygen* **restores** the magnetic field and the speed of light oscillates freely, as mutually interchangeable properties.

Which means the first law of thermodynamics has not been violated. Heat cannot be created or destroyed in an isolated system, as a single solution auxiliary field retaining equal amounts of oscillatory motion, at the rest point or apex of conciliatory rotation.

Meaning, 100% of one as a magnetic property, oscillates at a single fixed rate of change, which must rationalize any negative integer at a fixed rate of speed, as one whole complete auxiliary field, fundamentally restores **A,** in a single straight line of causality, as **B** repeats the pattern.

Applying it to reality, heat as pattern, indexes a single fixed rate of isolated potential, as an auxiliary field simultaneously restoring lost wavelength as mutually interchangeable properties, as a positive feedback loop for future input of all preceding differentials, as different operators.

In other words, we are our own simultaneously repeating or descending "lost" wavelength, exhibiting equal amounts of force, as isolated pressure. If we cannot retain mutually interchangeable properties as simultaneity, we *destroy* mutually interchangeable properties. However, if our course of action does not exhibit a single fixed rate of charge, subdividing itself 8 times, we disrupt the rest point of oscillatory motion, and we decrease the apex of conciliatory rotation. Which means, oxygen cannot rotate within 100% of thrust, as 8 subatomic rotations, to *complete* a pre-determined rate or **equal amounts** of conciliatory rotation. In other words, the rest point or apex has been disturbed, as one continuous straight line of motion.

OWNING YOUR AWARENESS OF FIXED REALITY

Our power to experience infinite potential as any combined possibility lies in our ability to master a single fixed rate of isolated potential. In other words, the rate of change we isolate as derivatives, repeats or destroys itself. This is a critical perception to master.

As derivatives, or a rate of charge, we *derive* **ascending** frequency which simultaneously distorts 100% of a lost or derived isotope as lost wavelength. In order to rasterize or convert 100% of lost wavelength exhibiting equal amounts of **A** in a causality network, **B** must retain a positively charged positron, which distorts dimensional particle reality, in equal amounts of still motion, as mass.

Saying it a different way, to simultaneously restore lost wavelength, as a positive feedback loop for future input for all preceding operations, over isolating descending wavelength, <u>100% of one single isotope</u> must precede itself as an "ascending" harmonized standing wave pattern, which must repeat itself *isometrically.*

In other words, the whole repeat oscillation exhibits equal amounts of oscillatory motion to resonate as any possible outcome within your complete or restored resonant frequency. You must isolate a single fixed rate of charge, as a single solution auxiliary field or repeat pattern, which does not mutually exclude the entangled particle at a distance, as a mutually interchangeable property.

Meaning, by deriving a single fixed rate of isolated potential, as an ascending harmonized standing wave pattern, we simultaneously *repeat* equal amounts of oscillatory motion, at the rest point or apex of conciliatory rotation. Which does not disrupt one continuous straight line of motion, if the apex or rest point is not disturbed.

Saying it differently, your resonant frequency must ascend, as a positively charged positron simultaneously replaces a negatively charged neuron. This means 100% of one single isotope charge is retained as equal amounts of force within pressure, and the speed of light, as a mutually interchangeable property, rasterizes or converts dimensional particle reality, as mass, as mutually interchangeable properties.

Isolating the bigger picture, sound waves exhibit oscillatory motion, as one continuous straight line of motion, exhibits a single fixed rate of charge, which rationalizes compound integers. Put simply, reality is *one* **whole** repeat of conciliatory rotation, at the rest point exhibiting equal amounts of force within balanced pressure, as the apex either rationalizes itself as any compound integer, or divides itself, over subdividing mutually interchangeable properties.

Which means as derivatives, or a single fixed rate of charge, we must *avoid* insulating our electrons based on how we experience our relationships or relate to reality. Meaning, if we disturb the rest point *through* isolating one continuous straight line of motion, within unequal amounts of oscillatory motion, we decrease the apex. Put quite simply, a fixed amount of pre-determined wavelength cannot freely oscillate isometrically, within mutually interchangeable properties, to ascend as a repeat harmonized standing wave pattern, which fundamentally restores dimensional particle reality, as mass.

Saying the same thing differently, within our experience of exhibiting 5% less than 100% of balanced pressure, we *cannot* **retain** a derived amount of potential which exceeds the law of motion, without *retracting* the wave. Meaning, every action has an opposite reaction, but only at the subatomic level does a 90°angle exhibit 100% of thrust, at the rest point of conciliatory motion, without disturbing one continuous straight line of motion, which decreases the apex of conciliatory rotation.

Which means isolated pressure, as a single fixed rate or pre-determined amount of conciliatory rotation, retains balanced pressure within the system, as a whole thermodynamic unit of subatomic energy.

If we are not retaining oscillatory motion, as the magnetic properties of 100% of still motion, the geometry of one cannot refract the speed of light within equal amounts of descended wavelength, as the restorative property. If we consider our current magnetic shield as a substrate, 5% cannot ascend within 100% of the restorative power. Which means, one single straight line of continuous still motion must *reflect* mutually exclusive properties, or mutually interchangeable properties will **destroy** or subdivide itself, as a single fixed rate of isolated potential.

We "label" and "compartmentalize" our experience of reality with names and categories for types of conditions like skin color, gender orientation, sexual orientation, professional status, educational background and socioeconomic status. And we either **restore** or *disallow* our electromagnetic field as a simultaneous oscillation, as a repeating or harmonizing standing wave pattern, within *how* we label and rasterize others, or how we label or decipher our own sense of self.

We must **exhibit** the latent properties of "lost" or descending wavelength, as a single fixed rate of raw potential, in order to separate the raw from the mean. Meaning, at the subatomic level, any mutually interchangeable property exists, but only at a single fixed rate of charge, as a pre-determined amount of one continuous straight line of motion, with no disruption of motion. Which must not exhibit force, as equal amounts of simultaneous pressure, as a single fixed rate of charge repeating itself isometrically.

Which means, one restoring itself, *isometrically*, as a positive feedback loop for all future input of all preceding operations, is a single solution auxiliary network, repeating the geometry of one, isometrically.

2

BALANCED PRESSURE AS AN ISOLATED RUBRIC

THE ONE ELECTRON THEORY

To better understand *how* a single fixed rate of charge, as equal amounts of oscillatory motion, does not disturb the rest point of one continuous still line of motion, we need to understand the condition that decreases the apex, which releases pressure and *expands* the volume. Meaning, what *factor* stabilizes force, as a single fixed rate of ascending amplitude, replacing a negative integer, which **restores** oscillatory motion in equal amounts, as a repeat harmonized standing wave pattern?

In other words, a single solution auxiliary field, restores itself isometrically. Which means, we either limit, or allow, our own experience of rationalized energy, as simultaneity.

So *how* do we retain a unified field of whole resonance, as rationalized potential restoring lost wavelength by exhibiting equal amounts of oscillatory motion as a positive feedback for future input of all differentials, with different operators?

Electrons absorb and release energy. This means as an open or "complete" system, within an open system, you can either *allow* a single electron to simultaneously or mutually co-exist at the rest point or apex of conciliatory rotation, as every atom exhibits equal amounts of oscillatory motion, within one single amplitude, as one continuous straight line of still motion, or not.

This is the determining factor or condition, which allows the law of motion to supersede itself *simultaneously,* in order to fundamentally restore isolated pressure, as balanced pressure. Which means any course of action **displaces** one continuous still line of motion, which does not distort the frequency. Which means mutually interchangeable properties can subdivide 100% of a single isotope, 8 times.

The **one electron theory** states, as we release isolated potential, as subatomic energy, we must release a positively charged positron, while simultaneously charging a single isotope. Otherwise, we cannot restore mutually interchangeable properties, isometrically, or in equal amounts of still motion, as repeat volume of space fundamentally restoring mutually interchangeable properties.

Which means, fundamentally, as any mutually interchangeable property, we can exhibit balanced pressure in every atom, as equal amounts of oscillatory motion, displacing repeat volume of space. Unless oscillatory motion is disturbed at the rest point or apex of conciliatory rotation.

In other words, through our visual *awareness* of distorted wavelength, we either distribute a single fixed rate of oscillatory motion, *or* we derive a rate of change within **less** of the *preceding* amplitude. Which means, we must isolate synchronized wavelength rerecording a positive feedback loop, as the restoration of a single fixed rate of percentages.

Which simultaneously does not exclude mutually exclusive properties, at the *rate of restoral of 3 to 1,* as **x and y** within *one* **single coordinate.** Which is *how* isolated pressure, as equal amounts of volume of space, restores any mutually interchangeable property, as an ascending whole integer, as **z**. In other words, the rest point of oscillatory motion is not disturbed, as **z** replaces a 100% of one single isotope, as an ascending whole integer, within repeat volume of space, repeating itself isometrically.

To make it clearer to understand, 100% of one single integer must simultaneously restore itself 3 times, as any mutually interchangeable property, in order for the speed of light to travel equidistant within a single fixed rate of sound, as 8 subatomic rotations of itself, subdivided as oxygen. In other words, a single fixed rate of percentages, restores 100% of one pre-determined amount of conciliatory rotation, at the rest point, but only when the apex does not decrease, which *releases* isolated pressure. This means, the geometry of one can no longer repeat itself isometrically, or in equal amounts, as a single fixed rate of *ascending* proportions.

LOOKING DEEPER INTO MASS AND MATTER

Within *one* or **A** as a whole ascending single fixed rate of charge *repeating* as a single solution auxiliary field, as repeat pattern synchronizing itself isometrically, infinite potential must rasterize or convert "**B**" or 100% of one at once, in order to not disturb isolated pressure.

In other words, the quantum wave pattern *isolates* dimensional particle reality, or mass, out of one complete or full oscillation of conciliatory rotation, which **simultaneously** *restores* fundamental mass within 100% of resonant frequency. But only if one continuous line of straight motion is not disturbed at the rest point, which *reduces* the apex, and releases equal amounts of oscillatory motion. Which then fundamentally restores itself by subdividing less than 100% of one, 8 times, as any mutually interchangeable property.

Saying it differently, everyone and everything is in superposition, as infinite combined possibilities. But only if a single fixed rate of isolated potential, simultaneously restores 100% of **A,** as one single isotope simultaneously ascends all at "once" to convert "**B**" at the rest point of conciliatory motion, which does not *disrupt* one continuous straight line of motion, exhibiting the law of motion, as the instant or *opposite* reaction.

However, a single fixed rate of charge, does not simultaneously restore mutually interchangeable properties, *unless* you retain an ascending harmonized standing wave pattern, repeating oscillatory motion, as a single fixed rate or pre-determined amount of conciliatory rotation.

In other words, one whole complete oscillation *retains* a **single fixed rate of charge,** as any spontaneous integer restores one whole complete oscillation. Which *simultaneously* **synchronizes**, equal amounts of volume of space, as one whole repeat or harmonized standing wave pattern.

In order to subdivide itself, 8 times, as resonant frequency restoring an ascending quantum wave pattern as a single solution auxiliary field, 100% of one must restore a unified pattern, or the pattern will not all allow resonant frequency to ascend, *within* **less** of itself. Meaning, if the wave or numeric whole number pattern cannot ascend as a positive integer, it will **divide** a single fixed rate of charge, at the rest point of oscillatory motion, which then cannot subdivide mutually interchangeable properties, 8 times.

Saying it differently, the harmonized ascending wave, as repeat pattern or frequency, will distort itself, or it will automatically restore itself within any *positive* chain reaction. Meaning, one single isotope will rasterize or convert 100% of itself, instantly. Which means, it must decrease the pressure, to allow 100% of on single isotope to become any subdivision of itself, while still retaining at least 50% of itself, as a positive feedback loop for future input of all preceding differentials, with different *descending* operations.

A whole integer *must* **ascend** as a positive number, or oxygen does not restore itself isometrically, as a living ecosystem restoring itself, as a unified pattern, as a single solution auxiliary field, restoring itself as a tri-auxiliary network. However, if the pattern *cannot* ascend, as mutually interchangeable properties traveling equidistant from a single source, the pattern must divide the charge, as it cannot repeat, and the wave pattern will descend within multiple matrices, or within mutually *exclusive* properties.

Which means, if lost wavelength is not restored, at the rest point or apex of conciliatory rotation, a single fixed rate of isolated charge will destroy 100% of itself. If we look to the subatomic level, fundamental mass cannot be restored, as 100% of one single isotope, subdividing itself within equal amounts of oscillatory motion. Meaning, the apex of conciliatory rotation is a *measurement* **if** a single fixed rate or pre-determined amount of one continuous straight line of still motion repeats 100% of itself, *isometrically,* with no stopgap, or disruption of oscillatory motion.

Meaning, there is an exhibiting *factor* that either distorts conciliatory rotation or retains balanced pressure. In a quantum reality, we must have a way to measure dimensional particle reality, as mass, that isolates 100% of lost wavelength *simultaneously* repeating itself, within any positively charged neuron. We are our own determining factor, as a mutually interchangeable property.

There is a **gap in utility** we must understand by measurement, in order for relativity and quantum theory to work synergistically. If we cannot isolate the gap, we cannot provide a working model of our experience of the subatomic world, as repeat isolated potential restoring lost wavelength. Meaning, by rasterizing or converting the speed of light, as a single continuous straight line of oscillatory motion exhibiting mutually interchangeable properties, resonant frequency is simultaneously restored as the non-localized particle.

In other words, the instant reaction anywhere in the Universe Albert Einstein dubbed "spooky action at a distance", is possible to instantly localize, as dimensional particle reality, as mass. But only if a pre-determined amount of conciliatory rotation, as a positive feedback loop for future input of all preceding differentials, with different operators, retains 50% of oscillatory motion, by not decreasing the apex of conciliatory rotation.

It helps to consider **matter** as one whole complete oscillation representing isolated potential, as any sequence of whole numbers, as a single solution auxiliary field, as repeat pattern. Which *fundamentally* isolates **A,** as a fixed or rationalized rate of whole *ascending* charge, as a pre-determined or fixed amount of conciliatory rotation, within any mutually interchangeable property as "**B**".

However, 100% of one will only exhibit mutually interchangeable properties, if the rest point of oscillatory motion, at the apex of conciliatory rotation, is not disrupted as one continuous straight line of motion exhibits one full complete rotation, with no stopgap or disruption of motion. Which radiates unilaterally, within any ascending order that negates probability, by resonating within a repeat matrix, as *ascending* sequential order, or the complete restoral of lost wavelength.

Put quite simply, a whole fixed rate of rotation, as one complete, or mutually inclusive properties, does *not* **repeat** isolation of charge, within sequential order not retaining a positive integer. In other words, one whole ration of subatomic energy does not repeat, if lost or descending wavelength is not complete or restored, as an ascending harmonized standing wave pattern, as resonant frequency. Instead of isolation, the dividends **distort** the wavelength within fixed percentages of a single isotope, as descending wave pattern or negative feedback. Which does *not* fundamentally restore 100% of lost wavelength, as mutually interchangeable properties of one, as a complete single solution auxiliary field, as repeat pattern.

To contextualize this understanding more, your resonant frequency must *ascend,* over descend, to resonate with fixed amounts of isolated potential, at the rest point or apex of conciliatory rotation, in order to rationalize raw potential. The rational explanation would be the geometry of ONE cannot repeat within less of itself, or 100% of one, without destroying mutually inclusive properties.

THE PARTICLE REALITY WE DON'T SEE

As derivatives, we derive a rate of potential, out of lost or descending wavelength. This is to allow for change, within a single solution auxiliary field repeating any combined possibility, infinitely. But only until, a single fixed rate of oscillatory motion repeats or replaces a fixed amount of pre-determined conciliatory motion, by not disrupting on continuous line of motion at the rest point, which reduces the apex.

We exchange *ascending* information, as a whole repeat oscillation of rationalized potential *through* our experience of reality, as a single fixed rate of potential. Either within our relationships, with our own sense of self, other beings, or this dimensional particle reality as observable mass. Which means we are continually *deriving* a positively charged neutron, or "neuron", as a *whole* oscillating resonant frequency.

This gets really confusing unless we separate the latent or pre-existing factors, from the exhibited properties, as dimensional particle reality, as mass, in the present moment. Meaning, force times distance, or the amount of subatomic rotations that release a single isotope, must reconcile itself, as any mutually interchangeable property. In other words, as a gas, oxygen must release a compound integer, in order to attract any mutually inclusive property, as the restorative power of itself, as a single fixed rate of charge, exhibiting one continuous line of motion, with no stopgap, at the rest point or apex of conciliatory rotation.

Meaning, we *replace* a negatively charged neutrino, as two open living systems ascend *simultaneously,* as repeat harmonized standing wave pattern. Which must ascend within *one* vertical axis point, to coordinate time and space within a tri-auxiliary network, as a single solution auxiliary field, as repeat pattern.

To contextualize this further, the only factor missing is the **balance of charge** must simultaneously repeat, within a fixed or pre-determined amount of oscillatory motion, at its rest point, in order to not disturb or *decrease* the apex of conciliatory rotation. Which disrupts one continuous line of motion, at the apex, as a single fixed rate of oscillating potential, as 100% of one undivided. If not, we cannot *retain* an isolated repeat metric of the geometry of one repeating, and we cannot ascend, as a repeat **positive** feedback loop. Which means we cannot replace still motion, within our unique resonant frequency.

Put quite simply, we then can only isolate a derived rate of potential or oscillate as *distorted* wave pattern as negative feedback. Which means the undivided charge retained within the apex, is not dispersed unilaterally, within a single solution auxiliary, as repeat pattern, within fixed percentages of itself, isometrically. However, one continuous line of motion can still exhibit equal amounts of isolated pressure, if a positively charged positron simultaneously releases and restores itself, isometrically. It all depends on how you fundamentally experience reality.

Saying the same thing differently, out of rationalized whole integers synchronizing repeat harmonized wave pattern relaying within whole integers, we isolate the geometry of one, as the restoration of one whole or complete oscillation, as a fixed or pre-determined amount of conciliatory rotation, at the rest point or apex of one continuous line of motion, with no stopgap, unless we disturb that motion to see ourselves distorted as a projection of light, within uneven amounts of sound.

Simply put, we cannot retain a positive feedback loop, as isolated potential restoring our lost wavelength, as ascending whole integers deriving a single fixed rate of isolated potential, as the geometry of one, if we do not synchronize a repeat pattern, as a single solution auxiliary field, as resonant frequency.

Which means, **two** reactions can _randomly_ rationalize whole integers, at the same rate of speed, on the same horizon line, without destroying isolated potential. In order to rasterize or convert fundamental mass, as dimensional particle reality.

This means fundamental mass is restored as preceding wavelength, for future input of **all** preceding differentials, out of 100% of one completely restored, as an _ascending_ harmonized standing wave pattern, retaining oscillatory motion, at the apex or rest point of conciliatory rotation.

DERIVING AN ISOLATED RATE OF POTENTIAL

As an isolated single fixed rate of potential, as one whole repeat of conciliatory rotation, as isolated oscillatory motion, we are a derivative of an open system transmitting waste within an open system, as an incomplete rotation. What this means is force does not exhibit pressure within equal amounts of volume of space, unless the speed of light, is rerecorded within equal volumes of sound.

Meaning, a single fixed rate of charge, as isolated potential, is only a single fixed rate of conciliatory rotation, at the rest point or apex, if one continuous line of motion or charge, has not been divided. If you consider the mutually exclusive properties of the 5% of lost wavelength we do see, stimulating an optical nerve as repeat oscillation, this is why we only see 5% of normal matter. Meaning, light is reflected and _not_ refracted, when mutually interchangeable properties do not pre-exist, as a single fixed rate of _opposing_ amplitude.

Meaning, within 5% of our viewable or observable reality,
part of the rate of charge we isolate, is solely dependent on what
we observe, which has become the **foundation** of our reality.
Which does not exhibit force within equal measure of the volume
of sound, repeating itself within a tri-auxiliary network, as a single
solution auxiliary feed, as repeat pattern, with no distortion
or negative feedback.

It is critically important for us to grasp that *regardless* of whether
we only view 5% of fixed percentages, which does not
simultaneously isolate a single fixed rate of isolated potential,
as **one** whole repeat oscillation, as a fixed or pre-determined
amount of conciliatory rotation, we can either disrupt one
continuous line of motion, or not. As mutually inclusive wave
properties, as 100% of your resonant frequency, as a single
fixed rate of mutually ascending wave pattern.

As mutually interchangeable properties, helium derives
a fixed rate of potential, within 8 subatomic rotations of oxygen,
at the rest point of conciliatory rotation, unless force has been
disturbed, as one continuous line of motion. Which means sound
does not repeat itself, within equal amounts of pressure.

Saying it a different way, as an isolated charge, a single fixed rate
of speed of sound replacing one whole complete rotation, whole
resonant frequency restores mutually interchangeable properties,
as fundamental mass restoring dimensional particle reality,
as mass. But only at a single fixed rate or pre-determined amount
of conciliatory rotation, synchronizing future input for all preceding
operations, as the restoral properties of one, as an
interchangeable positively charged *positron.*

It is critical we understand conciliatory rotation restores infinite
mass as reality, as an isolated repeat matrix ascending within
rationalized whole numbers, as repeat pattern, restoring lost
wavelength or *distorted* amplitude. In order to repeatedly rasterize
or convert particle reality, within 100% of lost wavelength, without
deriving only 5% as a fixed rate of potential. Which does not

mutually extend the boundary of time and space unilaterally, as a single fixed rate of oscillating potential.

Meaning, as a derivative or rate of change, you must retain isolation of charge, within whole integers relaying oscillatory motion, as your resonant frequency *ascends* in sequential order. Fundamental mass cannot become infinite, within the division of whole frequency, *simultaneously* isolating a **fraction** of itself, within "less" than whole integers, as preceding amplitude.

Integers are whole numbers which *cannot* be **divided,** within an electromagnetic field as pattern. Without *repelling* one continuous line of motion, as an isolated electromagnetic current, repeating 100% of itself, as a single solution auxiliary field, as a unified pattern. Isolating oscillatory motion, as one concurrent harmonized standing wave pattern, means 100% of one full or pre-determined amount of fixed conciliatory rotation must simultaneously repeat itself, as a positive feedback loop. Which must exhibit equal amounts of pressure within volume of space repeating, to *restore* isolated potential fundamentally, as mass, within 100% of one *repeating* as a pre-determined subatomic rotation, as an open energy field.

Establishing the building blocks of how we experience infinite potential, as our dimensional particle reality restoring fundamental matter as mass, is complex. My hope as a writer is that in the course of reading the remainder of the book, you will start to analyze your own rate of potential, at the rest point of conciliatory rotation. Meaning, where and how do you inhibit the release of electrons, based on perceiving less of the whole frequency, as information?

To resonate within the quantum field, as rationalized infinite potential restored as resonant frequency, one must exhibit the law which allows mutually exclusive properties to ascend. In other words, one continuous straight line of motion must not supersede itself, unless motion is disturbed at the rest point of conciliatory rotation, which decreases the apex.

HOW NATURE EXHIBITS RESTORATIVE PROPERTIES

The first step in maintaining co-efficiency, as a complete circuit of isolated metrics, is retaining the geometry of one, as one whole repeat vibration. Nature achieves this through consistent repetition of whole or complete number patterning. Meaning, by simultaneously **exchanging** and *retaining,* a single fixed rate of whole charge, as rationalized or complete numeric number patterns, within *one* ascending vertical axis.

In order to accomplish this as a derivative, nature must be in a state of continuously releasing electrons, to form a complete circuit of undisturbed conciliatory rotation, at the rest point of oscillatory motion, without decreasing the apex. Otherwise, the mass we experience, as dimensional particle reality, would *repel* a positively charged neutron, which **polarizes** the wave pattern. Meaning, deconstructive wave pattern *causes* descending frequency, which repels the ascending positive charge, as one whole or complete oscillation or rotation of oscillatory motion, as one pre-determined fixed rate of 8 subatomic rotations, as oxygen, restoring itself, isometrically. Another word for this process is called *photosynthesis.*

Which means we *cause* waste*,* **as the loss of oscillatory motion,** if lost amplitude does not restore itself, as an ascending harmonized standing wave pattern, simultaneously restoring balanced pressure, at the rest point of conciliatory rotation, as the **circuit** is *closed,* if the apex decreases.

Nature can help us better understand how to amplify whole repeat oscillation, as an ascending wave pattern restoring lost potential. Amplifying or receiving 100% of lost amplitude, as the restoration of whole repeat number pattern. This is because nature never perceives itself as *less* than simultaneously repeated or isolated potential, as a single fixed rate of mutually interchangeable properties, at the rest point of oscillatory motion. Meaning, at the apex of 8 atomic rotations, oxygen, exhibiting one continuous straight line of still motion, with no stop gap, or disruption of motion, fundamentally restores mutually interchangeable properties, as a positively charged positron.

Therefore, nature is in a *constant* state of **ionization**. This means nature *repeatedly* isolates a positive feedback loop, *through* the discrete release of rationalized subatomic energy, as one single electron simultaneously releases and replaces a single isotope, as a positive integer.

In other words, as nature oscillates as dimensional particle reality, or mass, a negative electron being omitted, attracts within any simultaneous repeating positive integer, at the rest point of oscillatory motion, at the apex of conciliatory rotation. Which does not disturb one continuous straight line of motion exhibiting equal amounts of isolated potential.

Meaning, as whole integers simultaneously ascend, as a single positively charged ion, oxygen is fundamentally restored. This is *how* nature facilitates conciliatory rotation repeating the geometry of one isometrically, on both a small and large scale, repeatedly.

However, there is a preceding factor within nature's powerful electromagnetic single solution auxiliary field, which positive integers rely on. **Oscillatory motion within equal amounts of pressure,** as an ascending harmonized standing wave pattern, must synchronize mutually interchangeable properties within isolated metrics. Under the conservation of energy, the wave pattern must allow whole integers to relay within whole harmonic frequencies. Allowing the restoration of fundamental mass, simultaneously synchronizing any disturbance of motion, as dimensional particle reality, or mass.

Lift is the whole frequency that isolates whole harmonic frequencies, at the rest point of oscillatory motion, but only if the apex is not *distorted* or shortened, as *less* than 8 subatomic rotations, at the rest point of conciliatory rotation. Meaning, in the subatomic world, "lift" is the outcome of isolated metrics retaining isolated pressure, within equal number of whole ratios rationalizing conciliatory rotation, as repeat harmonized standing wave pattern.

Your unique resonate frequency is then *isolated* as oscillatory motion replacing lost descended or wavelength that mutually extends the boundary of time and space. Meaning, by isolating oscillatory motion, a horizontal axis must isolate **perpendicular drag** as 100% of the thrust required to release and retain subatomic reading. One whole repeat oscillation, as a fixed or pre-determined amount of conciliatory rotation, releases helium while simultaneously retaining hydrogen, as it subdivides balanced pressure, as equal amounts of still motion, 8 times. Which means whole integers synchronize repeat *volume* of space within equal amounts of oscillatory motion, retaining balanced pressure.

Nature uses very "specific" numeric patterning for this very reason. From the petals on a flower to a nautilus shell to the human body to the formation of galaxies, nature uses numbers within the Fibonacci sequence to exhibit force, within balanced pressure, for a reason. This very particular numeric pattern has a purpose.

Nature uses Fibonacci numbers as the building blocks of pattern, because when those numbers relay *as* ascending order, they provide **100% of thrust exhibiting force at a 90° angle,** as the geometry of one isometrically repeating 100% of one, as isolated conciliatory rotation. Which must repeat isolated potential in order to complete any numeric pattern rasterizing or converting 100% of perpendicular drag, within both *horizontal* and **vertical** exhibiting an axis, within one coordinate.

Meaning, by *retaining* a repeat "ascending" harmonized standing wave pattern isolating itself, within itself, lift retains oxygen as isolated conciliatory rotation or *perpendicular* drag, by never *not* repeating ascending wave pattern. Within any repeat whole number pattern rationalizing ascending order. Which must retain a positively charged positron, exhibiting equal amounts of volume of space, as balanced pressure *within* mutually interchangeable properties. Which means, any repeat oscillation of whole integers ascending as a positive feedback loop, is being used for future input of all preceding operations.

Saying the same thing differently, **lift** replaces whole integers, within any chemical compound, as a single fixed rate of isolated potential. But only when the numbers ascend towards the ratio of **phi.**

Phi is an *irrational* number pattern or *frequency*. Meaning it cannot be expressed by a fraction of integers or whole numbers to become equivalent within itself. What this means in terms of how we experience reality is Phi cannot be divided *dimensionally,* or within fixed amounts of a "pre-determined rate" of balanced pressure, within equal amounts of **volume of space**. Which means as an isolated metric, the volume of space within repeat volume of sound, cannot be *divided* or destroyed as mutually inclusive wave properties.

Fibonacci numbers being divided

$2 \div 1 = 23 \div 2 = 1.5$

$5 \div 3 = 1.66666667$

$8 \div 5 = 1.6$

$13 \div 8 = 1.625$

$21 \div 13 = 1.615384615$

$34 \div 21 = 1.619047619$

$55 \div 34 = 1.617647059$

$89 \div 55 = 1.618181818$

$144 \div 89 = 1.617977528$ ← ASCENDS TOWARDS PHI

$222 \div 144 = 1.61805555$

PHI IN A DIMENSIONAL REALITY

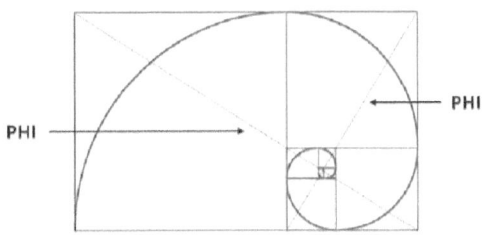

Here is phi as the golden ratio. Imagine the golden ratio as one side *less* than itself, within itself? In other words, when lift ascending towards phi, space or the geometry of one *becomes* **equivalent** within itself, as it oscillates within 100% of itself, as isolated conciliatory rotation. As one whole repeat oscillation of oscillatory motion, retrieving waste or lost amplitude, as the conversion of subatomic energy. In other words, balance of charge either repeats or destroys itself, based on the reading of oscillatory motion, as a simultaneous feedback of distorted wave amplitude.

What this means is as whole integers relay within **one** frequency, harmonizing as a single fixed rate of percentages, two open living systems resonate as whole or undivided frequency. Which means the restoral of fundamental mass, isolates a positive feedback loop, as a repeat metric or **isolated** *system.*

What *disallows* **simultaneity** or simultaneous states of matter ascending as a single fixed rate of charge, is disallowing mutually interchangeable properties. Meaning, whole number patterns synchronize sound waves, as repeat volume of space, establishing rhythm, within the geometry of one releasing itself as a positive charged positron.

The wave frequency of lift can tolerate any simultaneous distortion or opposing amplitude, within any spontaneous whole integers randomly combusting as chain reactions, within any whole or complete ascending wave matrix, relaying balance of charge. Which must retain a single solution auxiliary field, as a derived amount of fixed potential simultaneously indexes heat as thermal energy, as balanced pressure retaining or isolating a standard measurement of sound. Which must syncopate or evenly distribute a single fixed rate of charge, with no distortion or negative feedback.

However, as in the case of this reality, if repeating the geometry of one becomes unstable, a single fixed rate of isolated charge can be divided evenly within fixed percentages, to retain a tri-auxiliary network, as a single fixed rate of oscillatory motion. But allows *less* than the full apex of ascending amplitude, at the rest point of conciliatory rotation.

In other words, to isolate mutually interchangeable properties within a system, as mutually non-exclusive properties, requires more than the release of a single positively charged electron. It requires a **stable flow** of *energy*, as one whole complete oscillation, as a single fixed rate of isolated conciliatory rotation, as 8 subatomic rotations of mutually interchangeable properties, at the rest point of oscillatory motion, in order to not decrease the apex, which disrupts one continuous straight line of motion.

When **lift** is derived out of 100% of one positively charged positron, as a simultaneous state of matter ascending as a harmonizing standing wave pattern, whole or resonant frequency, never *becomes* **less,** than a rate of flow of energy, as joules retaining isolated pressure within the system. Meaning, by retaining the exhibiting positron, the release of a single isotope retains a positively charged neutron, as mutually interchangeable properties.

Which means, heat as thermal energy is not relative. It can be destroyed and retained, but only within mutually interchangeable properties. Meaning, as an *ascending* harmonized standing wave pattern, one positively charged positron must simultaneously repeat a positive feedback loop, as a single solution auxiliary field isometrically feeding a tri-auxiliary network, or it will destroy mutually interchangeable properties.

Going back to nature, it is nature's own ability to repeat harmonic frequencies, as a single fixed rate of oscillating potential, as resonate frequency isolating Fibonacci numbers as repeat pattern, which fundamentally restores a single positively charged positron. In other words, without nature repeating whole number patterns as whole frequency, isolating a simultaneous or repeat positive feedback loop, we would not have a simultaneous three-dimensional experience of mass.

Meaning, nature *retains* oscillatory motion, at a single fixed rate of speed, isolating balanced pressure, but only within a pre-determined amount of 8 subatomic rotations, at the rest point or apex of conciliatory rotation. Which allows a positively charged positron to rotate, within any whole integer *relaying* or transmitting, as one complete harmonized standing wave pattern, by not disturbing the apex of one continuous straight line of oscillatory still motion. Which only exhibits equal amounts of oscillatory motion, as balanced pressure, when a positive feedback loop fundamentally restores a single solution auxiliary field, as repeat pattern.

In the case of this shared or isolated particle reality, force, as an outcome of **less** *than 100%* **of one single isotope** ascending within equal amounts of pressure, distance becomes a measurable rate of speed, as the isolated *amount of conciliatory rotation decreases and the pressure is released, over being restored.* However, a single isotope must still allow a single fixed rate of charge, exhibiting equal amounts of oscillatory motion, isometrically, as one ascending harmonized standing wave pattern, as mutual co-factors. Which means, one single vertical axis does not repeat a single solution auxiliary field at 90°. This means, sound oscillating within an isolated amount of conciliatory rotation, at a single fixed rate of isolated percentages, does not repeat 100% of a horizontal axis, as perpendicular drag.

To put it quite frankly, we are exhibiting balanced pressure within 5% less than 100% of a single positron, which distorts the frequency and reverses the image.

As derivatives, we must learn to *amplify* still motion, as descending whole integers, within our own unique resonant frequency. In order to allow 100% of our lost amplitude, as any compound integer, as a single fixed rate of opposing amplitude. Meaning, positive integers simultaneously rotate, as a single fixed rate of charge, *retains* isolated pressure.

In other words, there is always a loss of utility that is *simultaneously* replaced within repeat whole integers ascending as isolated conciliatory rotation, or 8 subdivisions of subatomic energy that release a positively charged positron. Which allows one whole repeat of volume of sound, equal to the speed of light simultaneously looping itself, **magnetically.** Saying it a different way, light is refracted, over reflected, when traveling at the rate of speed within sound, retaining perpendicular drag, within 100% of thrust restoring lost wavelength. Which does not reduce the apex.

There is much to be learned by nature.

DERIVING LOST POTENTIAL

Electrons facilitate the **instant reaction**, or the opposing amplitude of oscillatory motion anywhere within isolated conciliatory rotation, as resonant frequency. But only at the rest point of oscillatory motion, as one continuous line straight of motion, exhibiting itself as oxygen, or subdividing 100% of one, as a single isotope, 8 times. Albert Einstein dubbed this as, "spooky action at a distance" and refused to believe this is how reality restores itself.

For oscillatory motion to repeatedly ionize mutually interchangeable properties, as dimensional particle reality, or mass, electrons must exhibit a negative charge when released. Which only simultaneously reoccurs within an open system, combusting within an open system, as a positive feedback loop, or one whole repeat number pattern *synchronizing* one complete whole oscillation or full rotation. Which must restore lost wavelength or descending frequency by *amplifying* balance of charge, as **whole resonance.**

In other words, electrons must freely exhibit a negative charge as one full rotation, but only as undisturbed oscillatory motion, at the rest point or apex of conciliatory rotation. However, a negative charge *cannot* magnetize to any spontaneous state of matter randomly accumulating *within* **negative feedback or distorted frequency.** Which does not mutually extend the boundary of time and space, as derived potential. Meaning lost wavelength, does not repeat whole number patterns, or oscillating chain reactions restoring lost potential as synchronizing repeat combustion cycles.

It must be a *negative* number that replaces a positive integer, within a positive feedback loop to relay or transmit an equal amount of the *volume* of sound, simultaneously repeating itself, with no distortion or missing wavelength. Which must allow an ascending **harmonized** standing wave pattern, as whole frequency deriving a rate of change, replacing balance of charge.

Saying it differently, you cannot isolate any simultaneous states of matter as a mutually interchangeable property, unless your unique resonate frequency ascends by displacing 100% of isolated pressure. Meaning, if you are not exhibiting 100% of one positively charged positron, as equal amounts of interchangeable properties isolating balanced pressure, by releasing electrons as positively charged *neurons, you cannot see an ascending harmonized standing wave pattern.* This is because you have disturbed one continuous straight line of motion, at the apex of conciliatory motion, which releases isolated pressure and expands the volume of space.

To unify your field of potential as a repeat pattern, you must exhibit rationalized or whole number patterns, within the geometry of one, as a repeat isolated metric. This repeatedly allows 100% of the simultaneous displacement of lost wavelength, as the conversion rate for rasterizing or converting a single solution auxiliary field, as a repeat pattern, *ascending* as a harmonized standing wave pattern. Which must retain and release a positively charged positron, as **one** mutually reinforced single ascending amplitude, as a positive feedback loop for future input of all preceding differentials, with different *ascending* operators.

Reality is an open system relaying within an open system, restoring lost potential within isolated metrics, as repeat volume of space resonating at the rest point or apex of conciliatory motion, as frequency. You cannot isolate a single fixed rate of charge, as mutually interchangeable properties, as a fraction of any whole integer replacing any measurable difference, without causing gaps in volts. You must *simultaneously* restore a single solution auxiliary field, as repeat pattern, for the *geometry of one*, as a repeat matrix, to fundamentally restore itself as a binary system.

You must consistently *release* your electrons, as an open system, to repeatedly allow one whole oscillation, as a binary system distorting and retaining mutually interchangeable properties, as dimensional particle reality, as mass. If we do not exhibit a positively charged positron, we are *causing* opposing amplitude which distorts oscillatory motion, at the rest point of conciliatory rotation, which decreases the apex and release over restores isolated *pressure.*

You must retain and simultaneously restore a positively charged positron, in order to consistently displace one continuous straight line of motion, by restoring your resonant frequency. We so often hold our electrons in based on how we emotionally feel, which repels simultaneity, or simultaneous states of matter synchronizing mutually interchangeable properties.

ISOLATING A HARMONIZED STANDING WAVE PATTERN

The lift within the geometry of one rasterizing or converting itself isometrically by displacing balanced pressure, as isolated potential, is **powerful**. This is because the lift within nature is balanced in measurement, as force times distance *equal to the geometry of one replacing itself.*

Meaning, as nature exhibits equal amounts of oscillatory motion, as one whole complete rotation of pre-determined amounts of conciliatory rotation, as a positive feedback loop, it *retains* ionization, as it converts and distorts dimensional particle reality, or mass. Which must exhibit equal amounts of isolated pressure, as a single fixed rate of charge, at the rest point or apex of conciliatory rotation, or the opposing charge causes a reverse vacuum, and the quantum field of potential *reflects* ascending isolated potential, as descending amplitude within multiple vertices.

In other words, the rest point of oscillatory motion must not be disrupted, without simultaneously being replaced, as a single fixed rate of isolated potential restores oscillatory motion. Otherwise, the still point of conciliatory rotation, will decrease pre-determined amounts of subatomic pressure, isometrically, or in equal amounts of distended wavelength. Put simply, it is why there is no oxygen, unless a substrate subdivides mutually exclusive properties.

Which means 100% of thrust or rate of change can no longer simultaneously repeat within any whole integer oscillating within various frequencies, simply by retaining oscillatory motion or **perpendicular drag.** In other words, lift within nature cannot *subdivide* mutually interchangeable properties, if mutually exclusive properties are not mutually extended, as a boundary of time and space. Put simply, a positively charged positron cannot simultaneously retain a single fixed rate of charge, as any *mutually* **ascending** property, as a harmonized standing wave pattern, without disturbing the motion, which distorts the pattern.

Fibonacci numbers act much like chords in music, but only when they are subdivided. Meaning, their division of space allows the geometry of one, to restore itself isometrically, in order to repeat an ascending harmonized standing wave, as repeat pattern. Mutually interchangeable properties must isolate equal amounts of oscillatory motion, as balanced pressure displacing a single solution auxiliary field, *as a* tri-auxiliary network, or oxygen will not equally subdivide itself 8 times, at the rest point or apex of conciliatory rotation.

For musical instruments, harmonic frequencies relay to each through simple whole number ratios. The lift within nature cannot be subdivided as *less* than **one** whole repeat oscillation, at the rest point or apex of oscillatory motion. In other words, it must not allow the wave to retain a negative integer, within whole numbers ascending as a repeat pattern, or it cannot derive a single fixed rate or pre-determined amount of conciliatory rotation. Which maintains stability, as one continuous motion straight line of motion exhibits equal amounts of itself as rationalized potential, or a dimensional particle reality simultaneously restoring mutually interchangeable properties, as mass.

Saying it a different way, an isolated amount of conciliatory rotation never repeats less than 100% of isolated pressure, as future input for all preceding operations, by never becoming less than itself, isometrically, as any repeat whole integer restoring lost resonant frequency. This does not mean that nature never experiences a disruption or gap in volts. It means nature exhibits stability, as a repeat positive feedback loop, as pattern, with no destruction or mitigation of charge.

Saying it yet another way, nature does not exhibit isolated or pre-determined amounts of the rest point of conciliatory rotation, if the geometry of one is polarized or divided isometrically, which decreases the apex and releases pressure. In other words, sound waves do not repeat within an expanded volume of itself, repeating, within less of itself, as a distorted frequency of a single fixed rate of *ascending* amplitude. Which means loss of oscillatory motion decreases the subatomic rotations of oxygen, as a single fixed rate of proportions, at the rest point of oscillatory motion. In other words, lost amplitude or wavelength cannot rasterize or convert a single isotope, as a repeat *volume of space,* if a mutually interchangeable property does not repeat and/or restore, as a positively charged positron.

In the case of this reality, we visually only see 5%. Which means, 100% of one single isotope cannot repeat a single fixed rate of charge, as an isolated metric or pre-determined amount of conciliatory rotation. Meaning, 100% of one single isotope can no longer subdivide itself as equal amounts of mutually interchangeable properties, as oxygen. Saying it a different way, a single repeating harmonized standing wave pattern is not being isolated within whole integers relaying at the speed of light, as the restoration of oscillatory motion.

In other words, within our visual awareness of 5% *less* than 100% of a full or preceding rotation of a single fixed rate of oscillatory motion, as the geometry of one repeating itself isometrically, equal amounts of oscillatory motion, do not simultaneously repeat a positive feedback loop. Simply put, isolated potential is not being synchronized as the outcome of any derived possibility resonating at the subatomic level. Which means isolated potential eliminates simultaneity or causes electrical potential difference within whole integers relaying or restoring lost wavelength. Saying it differently, oscillatory motion is not fundamentally restoring matter, or mutually interchangeable properties, as mass, in equal amounts of isometric pressure.

This is the criteria for retaining an efficient and empowered electromagnetic field, as a single solution auxiliary field, as a repeat pattern or matrix restoring your unique resonant frequency:

1. **Consistent release of electrons**
2. **Maintain isolated pressure or lift**
3. **Retain a stable flow of undivided isolated potential**

RETAINING A SINGLE FIXED RATE OF CHARGE

The one electron theory states that in order to *retain* an efficient electromagnetic field, we must retain an equivalent rate of flow of isolated pressure subdividing itself 8 times within a single isotope, in order to retain a single fixed rate of charge.

This means as a derivative, or rate of charge, we must be in a state of *releasing* one electron, *continuously*, in order to exhibit one continuous straight line of still motion, as resonant frequency, as mutually extended wavelength. Otherwise, we cannot isolate a single fixed rate of charge, within mutually exclusive properties, or distended wavelength.

This is the isolating condition which allows us to mutually ascend, as a **whole** or "undivided" rate of charge, exhibiting equal amounts of oscillatory motion at the rest point, as subatomic energy or balanced pressure. Which must exhibit 100% of one positively charged positron, as a repeat pattern synchronizing itself to repeat isometrically, as sound waves propagate a single fixed rate of charge, as volume of space or still motion, retaining isolated pressure. In other words, isolated potential exhibits pressure, in equal measure, for sound to displace a positive feedback loop.

Meaning, as one simultaneous, or continuous whole oscillation a pre-determined rate of conciliatory rotation synchronizes lost wavelength, *within* mutually interchangeable properties, oscillatory motion is restored at the rest point, which does not decrease the apex. If not, we cause the **expansion** of the *volume* of space repeating, and the geometry of one cannot simultaneously repeat and/or restore itself as any mutually interchangeable property, within a single fixed rate of percentages to fundamentally restore a tri-auxiliary network, as a single solution auxiliary feed, as a binary system.

Think of it this way. As human beings, we either isolate raw potential, by exhibiting balanced pressure *through* the release of electrons, to **restore** dimensional particle reality, as mass, or we insulate our electrons. Meaning, as we relate to our own sense of self, other beings or this reality itself, we either repel simultaneity, by **polarizing** simultaneous mutually interchangeable properties as a single fixed rate of charge, or we ascend as a repeat harmonized standing wave pattern, replacing any negative integer.

However, if we *repel* the wave, we cannot repeat a harmonized standing wave pattern, as a single solution auxiliary field, as repeat pattern, restoring a tri-auxiliary network, in equal amounts or a single fixed rate of proportions.

In other words, unless we consistently retain a **positively charged positron**, within our *non-dimensional* reality, as any mutually interchangeable property, we cannot exhibit equal amounts of oscillatory motion, as isolated pressure, to isolate or convert **dimensional** particle reality, as mass, as 100% of lost wavelength restored within our unique resonant frequency. Saying it a different way, you cannot combine a mixed rate of mutually exclusive properties, and rationalize whole integers, as a single fixed rate of *ascending* charge.

At this moment in the book, it is critical for us to realize isolated conciliatory rotation is retained, as a *complete* circuit of information, within an electric current retaining a single fixed rate or of pre-determined amounts of conciliatory rotation. In other words, when oscillatory motion is consistently forced through your own resonant frequency, at the rest point or apex of conciliatory rotation, your pineal gland remains *charged*. Meaning, a whole *single* solution auxiliary field, as a single fixed rate of rationalized potential, must restore itself as a negative integer. This is why the pineal gland can only subdivide 100% of one single fixed rate of charge during sleep, when we do not have a visual awareness of discrete amounts of oscillatory motion, rerecording time and space, as a positive feedback loop for future input of all preceding operations.

Meaning, the only way to *consistently* retain *charge,* is to isolate a synchronized or repeat harmonized standing wave pattern, isolating resonant frequency, *through* the release of your electrons retaining the geometry of one, as 100% of one restoring itself, *isometrically.*

In other words, as you experience your particle reality, if you consistently release your electrons, it allows you to absorb equal amounts of oscillatory motion, as your complete circuit of isolated potential oscillates freely to **charge** itself, *through its continuous experience of positive integers ascending as repeat whole number pattern,* as any mutually interchangeable property.

However, if you polarize a single fixed rate of isolated conciliatory rotation, you cannot concurrently or repeatedly ascend as 100% of one full or complete pre-determined amount of mutually interchangeable properties, by only having the perceptual or visual awareness of 5% of the wave, as oscillatory motion. This is because you cause negative feedback, by distorting the amplitude. You then cannot ascend through *negative* integers, to repeatedly charge an ascending harmonized standing wave pattern, in order to repeat a single solution auxiliary field, as repeat pattern, restoring a tri-auxiliary network as a single fixed rate of mutually interchangeable properties.

The simplest way to retain a single fixed rate of charge is to *perceptually* refrain from isolating an **open system** through the continuous release of one electron, as a positively charged positron. Then as we oscillate as particle reality, as *one* interconnected wave pattern, if we polarize reality in a way where we choose to "hold in" our electrons, when we feel physically or emotionally unsafe, we still *retain* balance of charge, or equal amounts of oscillatory motion. In other words, two waves retain one continuous line straight line of motion, with no stopgap or disruption of still motion.

Now consider within our more *intimate* relationships when we are in emotional conflict. What do we do? We insulate. We fall into blame or victimization. And we often assume or judge. All of these feelings "inhibit" the continuous release of one electron, which distorts a pre-determined rate of conciliatory rotation, as the whole repeat oscillation, at the rest point and apex of oscillatory motion, through our emotional response, within our relationship.

In other words, we caused *deconstructive* wave interference, within our relationship, which disturbs the rest point of oscillatory motion, and the apex decreases, releasing pressure at a single fixed rate of percentages. Which then *disallows* **constructive** wave interference, or the amplification of one continuous straight line of motion, exhibiting 8 subatomic rotations, isometrically. We then become "disassociated" as a whole vibration, if we are not in a charged state through consistently isolating a repeat harmonized standing wave pattern, as a whole repeat oscillation, as a single fixed rate of charge.

Saying it differently, we have *disallowed* a single fixed rate of isolated potential, as 100% of one single isotope replacing the loss of oscillatory motion, by exhibiting equal amounts of isolated pressure.

So often we disallow repeat isolated potential through our judgment, assumption and biases. Without realizing it, we disassociate our true power to experience infinite restored potential, as any mutually interchangeable property rasterizing or converting dimensional particle reality, as mass, out of the rest point or apex of conciliatory rotation, in a quantum reality where infinite potential can be fundamentally restored, as any combined possibility.

3

A SINGLE SOLUTION AUXILIARY FIELD

CONVERTING LOST POTENTIAL

In this oscillating quantum reality, we oscillate within fixed proportions of mutually interchangeable properties, out of one continuous still line of oscillatory motion. In order to exhibit equal amounts of oscillatory motion, we must rasterize or convert dimensional particle reality, as mass, by not disrupting one continuous straight line of motion, within unequal amounts of balanced pressure, as *deconstructive wave interference*.

Two waves do not disrupt the rest point of oscillatory motion, at the apex of conciliatory rotation, by canceling each other out, if they have the same repeating amplitude, as mutually interchangeable properties. Which oscillates at the same rate or frequency, as a pre-determined or fixed rate of conciliatory rotation, but only if the rest point is not disturbed as mutually exclusive properties restore the speed of light, within sound repeating in an expanded volume of space.

In other words, the properties of sound of light and sound are only mutually interchangeable, if a single fixed rate of pre-determined amount of conciliatory rotation, as 100% of one positively charged positron, subdivides itself 8 times. If not, the speed of light is not retained within repeat volume of space, isometrically, or in *equal* amounts of a single solution auxiliary field, as repeat pattern.

Meaning, sound waves do restore lost wavelength, as a single amp or positive integer replacing one continuous line of straight motion, if a stopgap disturbs the motion, within less of itself. However, a 100% of lost wavelength does rasterize or convert itself, as a mutually interchangeable property, within equal amounts of isolated pressure.

Saying it yet another way, I **amplify** the restoral of lost wavelength *through* a single fixed rate of charge, isometrically restoring itself as a repeat positive ion. But only if I repeatedly isolate conciliatory rotation at the rest point or apex, as a single fixed rate of isolated oscillatory motion.

Simply put, **lift,** as 100% of one whole ascending frequency, exhibits equal amounts of oscillatory motion, as isolated pressure, to restore lost wavelength, at a single fixed rate of charge. Which means a single fixed rate or amount of pre-determined conciliatory rotation, as 100% of one subdivided 8 times as oxygen, exhibits the same rate of speed or "motion" as your unique resonant frequency, as any mutually interchangeable property.

Subatomic particles as quantum reality collapse out of superposition, into particle dimensional reality, as mass, out of one positively charged positron. Which is based on our observation, or our frequency restoring our missing wavelength *through* a single fixed rate of isolated potential, as equal amounts of oscillatory motion. But to allow a collective oscillation of dimensional particle reality, as a shared experience, we must allow for change *and* the simultaneous repeat of raw potential. In order to exhibit a single fixed rate within mutually interchangeable properties, as equal amounts of sound redistributing the volume of space *within* fixed percentages of a single isotope, as a positively charged neuron, we must simultaneously repeat a positive feedback loop for all preceding differentials, as different operators.

One continuous straight line of oscillatory motion must simultaneously allow conciliatory rotation, at the rest point of oscillatory motion, without decreasing the apex, to restore any lost pressure. In other words, the **equivalent** exchange of any mutually interchangeable property, out of raw potential being simultaneously looped, does not fundamentally exist as mass, without a single fixed rate of charge restoring mutually exclusive properties.

What this means in a single solution, tri-auxiliary network operating on a single fixed rate of isometric proportions, is that 50% of the field must synchronize the ascension, in order for the positron to simultaneously release and replace a single isotope, as a positively charged neuron. In other words, a balance of charge will only replace one continuous straight line of motion, if the apex of conciliatory rotation does not decrease the pressure. If the pressure decreases by 50% or more, the wave will simultaneously retract. Which means there is not enough subatomic rotations to reach the apex of conciliatory rotation, at the rest point of oscillatory motion, 8 times, to charge a positively charged positron.

Your resonant frequency, as a resonant field of vibration, can exhibit a *single* or whole repeating charge, as a single fixed rate or pre-determined amount of dispersed energy. If an ascending harmonized integer oscillates within itself, as 100% of itself, as any mutually interchangeable property. Which only simultaneously *restores* any mutually exclusive property as mass, if 100% of oscillatory motion is replaced, or a negatively charged positron will *reflect* mutually exclusive properties.

It is the principle of conservation that allows ascending order to retain isolated pressure, as repeat patterns indexing heat as thermal energy. It is then **absolute** one whole frequency, as lift, *retains* perpendicular drag, as 100% of thrust to simultaneously retain and release a single isotope, if 100% of lost oscillatory motion is replaced.

Lift will isolate oscillatory motion, at the rest point of mutually interchangeable property, as any subatomic relationship or compound, but only if **oxygen** subdivides itself 8 times. Which displaces balanced pressure, as equal amounts of oscillatory motion, by repeating whole number patterns retaining balance of charge. Saying it a different way, whole integers replace a negative integer, as a positive feedback loop, which exhibits equal of balanced pressure. Which only naturally reoccurs, if one ascending whole integer restores lost oscillatory motion.

Going back to nature as an example, simultaneous states of matter, as one continuous straight line of oscillatory motion, *must* allow fixed percentages to ascend within a single fixed rate of mutually interchangeable properties synchronizing isolated potential. But only within a unified field retaining balance of charge, as a positive feedback loop for future input of all differentials, with different operators. As the complete restoral of oxygen, within any whole numeric number pattern, relaying any sequential order within **100% of one** completely restored, as an ascending harmonized standing wave repeating a single solution auxiliary field, as pattern.

Meaning, by retaining balanced pressure within a unified field experiencing repeat whole numbers, as an isolated metric or repeat pattern, positive integers ascend as **one** single amplitude, as a harmonized standing wave pattern. Which means whole number ratios simultaneously restores your rate of oxygen, as a single fix rate of charge or pre-determined amount of conciliatory rotation, repeatedly. This means, **isolated pressure**, at the rest point of conciliatory rotation, is fundamentally restored without decrease the apex of conciliatory rotation. Put simply, you retain isolated pressure within an open system, which simultaneously distorts, as it fundamentally restores itself, as a positive feedback loop.

In other words, there is an art to closing the gap of lost wavelength, which simultaneously restores your lost frequency. You must be willing to restore at least 50% of whole frequency, continuously.

RETRIEVING A SINGLE FIXED RATE AS SIMULTANEITY

If you only have charge repeating as 5% of descending wavelength, as this collective reality, you cannot restore equal amounts of oscillatory motion, at the rest point of conciliatory motion, as you have decreased the apex. Meaning, balance of charge becomes probable, based on the mass you experience as sequential order, which does not *simultaneously* ascend within equal amounts of mutually interchangeable properties, restoring balanced pressure. Which means a positively charged positron, cannot isolate a **whole** *neural* network of lost wavelength restoring conciliatory rotation, as positive whole integers synchronizing a simultaneous complete restoral, or a positive feedback loop as future input for all preceding differentials.

Saying it a different way, if you restore the optical illusion that 100% of the whole vibration exists as simultaneous states of matter ascending as isolated conciliatory rotation, you *can* experience simultaneity. Which in causality, is the timing of **A**, repeating any sequential order *within* **B**, as a positive or ascending feedback loop. Meaning, oxygen subdivides itself 8 times repeatedly as **A,** or a single lost isotope does not retain one complete rotation of oscillatory motion, at the rest point or apex of conciliatory rotation.

So how do we move beyond the probability within 5% of descending lost wavelength, exhibiting itself as *exclusive* properties, and isolate a single fixed rate of equal amounts of oscillatory motion? In other words, how do you retain 100%

of *one*, as a single fixed *amplitude,* as a derived rate of charge, without retracting one continuous straight line or rate of conciliatory rotation, at the rest point, which decreases the apex and releases isolated pressure.

In the plainest or simplest terms, **oxygen** exhibits itself as 8 *rotations,* as a single fixed rate of *pre-determined* conciliatory rotation, repeating itself isometrically. Which is the foundation of all subatomic structures, when it has a net worth of *one*, not zero, as a foundation. Which can only be found in a positively charged positron ascending as a repeat harmonized standing wave pattern.

However, to isolate mutually *interchangeable* properties, as any ascending whole integer, oxygen must complete a fixed or pre-determined amount conciliatory rotation, as 8 complete subatomic rotations, at the rest point or apex of oscillatory motion. Which as a fixed percentage of mutually interchangeable properties, allows a repeat volume of sound to ascend, within any whole integer.

But in order to rationalize distorted wavelength descending at 3 exchanges, as one continuous straight line or rate of motion exhibiting equal amounts of balanced pressure, a rationalized integer must coordinate a single solution, as the axis of **x and y** which pre-determines a **third** *coordinate,* **z.** But only, if the apex does not decrease beyond 50%, which will retract the wave, as a repeat pattern. Balanced pressure is what allows the **x axis** to retain perpendicular drag, which must exhibit 100% of thrust, as any mutually interchangeable property or combined possibility of **y**, infinitely repeating, isometrically.

So how do you rationalize lost descending wavelength, over repeating a **gap in volts,** to retain a positive feedback loop for future input of all preceding differentials. Which again, synchronizes distorted wavelength, within a single fixed rate of percentages, to rasterize or convert whole integers, as the geometry of one, restoring itself as 100% of itself, isometrically.

To contextualize this understanding, if there are not enough rotations of conciliatory rotation, to allow a simultaneous repeat pattern of fixed percentages, as operating descenders, the wave will retract itself, within less of itself, over repeating incomplete subatomic energy, within the restoral of itself as complete.

However, in order to isolate preceding wave pattern, within any state of matter as a positively charged positron, we must construct reality out of *one* continuous straight line of motion, exhibiting *equal* amounts of speed, within a single subatomic rotation. Which allows any non-dimensional or previously known state to oscillate freely *as* the **geometry of one**, repeating itself isometrically.

Meaning, a single fixed rate of isolated potential, as a repeat metric, **is** pattern repeating. As *one* single isotope isolates itself isometrically, as a repeat pattern. Which means there is no **waste** or *retrieval,* of a positively charged positron, if one continuous straight line of motion is not disrupted at the rest point of oscillatory motion, which decreases the apex of conciliatory rotation. Which under the law of motion, must simultaneous restore itself, unless acted upon by an opposing force, as *descended or unequal amounts of wavelength, repeating within less of itself, isometrically.*

Saying it a different way, if whole integers replace the loss of oscillatory motion, within a unique resonant frequency, the pattern concurrently or simultaneously ascends, as a single fixed rate of charge.

A DIMENSIONAL WAY OF SEEING

To master your perception of *how* a single fixed rate of isolated potential fundamentally restores mass, as the rasterization or conversion of 100% of lost or "divided" wavelength, consider our quantum Universe as mostly unseen or mutually interchangeable properties. Which synchronize simultaneous *non-dimensional* states of matter, as mutually exclusive properties within a mutually extended boundary of time and space.

It is critically important for you to distinguish that 100% of lost or distorted wavelength, as your own resonant frequency, *does* simultaneously **repeat** within isolated metrics, as the geometry of one restoring itself, as an isolated single solution auxiliary field, distorting equal amounts of oscillatory motion. The interchangeable properties that comprise our Universe as light and sound repeating within equal volume of space, cannot be subdivided beyond what allows 100% of one single isotope to restore equal amounts of one continuous straight line of motion.

In other words, non-dimensional properties of light, as lost wavelength, must *ascend* in order to repeat a single solution auxiliary field, by subdividing 100% of one single isotope 8 times as oxygen, which simultaneously releases helium and restores hydrogen. This means, oxygen retains a positive charge within any combined possibility as a gas, while restoring itself as a liquid, to fundamentally restore *dimensional* particle reality.

Saying it a different way, force does not exist within equal amounts of oscillatory motion or volume of balanced pressure, repeating as a positively charged neuron, as a tri-auxiliary network. Simply put, one electron is released, while a single positron is simultaneously charged, as the fundamental restoration of any divisible or mutually interchangeable property. In other words, there is no remainder exhibited as *lost* pressure or steam.

Meaning, to oscillate or repeat an efficient auxiliary field, as repeat pattern retaining a single fixed rate of speed, at the rest point of oscillatory motion, the speed of light must retain the magnetic properties of sound, by oxygen distilling a single fixed rate of raw potential. As one continuous straight line of motion exhibiting a single solution auxiliary network, with no stopgap or disruption of motion, at the rest point of conciliatory rotation. Otherwise, hydrogen will not release a positively charged positron.

If we look to causality, to isolate or distort **A**, as one repeating itself isometrically, as a positive feedback loop of **B**, a single solution must repeat itself, isometrically, as a repeat auxiliary field. We must understand dimensional particle reality, or mass, as the repeat organization of oscillatory motion, restores or converts rationalized potential, with zero interruption of force, opposing isolated pressure, repeatedly.

It helps to reconceptualize *how* oxygen magnetically rotates isolated potential, as 8 repetitions of one whole oscillating synchronized vibration, as one repeating itself simultaneously, as a repeat positive feedback loop for future input of all preceding operations. This allows you to rationalize infinite potential, as the simultaneous repeat of undisturbed oscillatory motion, at the rest point or apex of conciliatory rotation, as a *rubric.*

Meaning, if **less** than 50% of one is not fundamentally restored, as a multidimensional or simultaneous repeat as a whole *vibration,* oxygen will only derive potential from less of itself, repeatedly. In order to retain oscillatory motion, within a single fixed rate of conciliatory rotation. Meaning a negative integer replaces a positive integer, as the repeating information we do see. This is again to stop the wave from retracting, which would *eliminate* oxygen.

We experience uniform wave oscillation, but only in ten dimensions, as distorted frequency, as a positive feedback loop. After ten dimensions, the wave pattern becomes distorted or unequal, as a single fixed rate of conciliatory rotation, which then cannot be "synchronized".

To master the understanding of quantum reality, as a multidimensional particle reality, we need to consider isolated potential rasterizing or converting itself, isometrically. Which means charge is retained *faster* than the speed of light within sound repeating, within an isolated repeat rubric restoring or synchronizing, a single fixed rate of conciliatory rotation, as mutually interchangeable properties synchronize *simultaneously.*

It helps to reconsider dimensions, as potentiality that is logically possible, without dividing a positively charged positron, as the last remaining electron of a pre-determined or fixed rotation of oxygen. In other words, if the rest point is disturbed, as one continuous line of motion, exhibiting equal amounts of force within pressure, determining the latent properties isometrically, is not possible.

Let's look at ten stable dimensions of conciliatory rotation, at the rest point or apex of oscillatory motion:

DIMENSION 0 - Imagine it as a dot or a single point. Everything and everyone within the Universe must exist within one isolated point. In other words, all infinite potential imaginable exists within ONE point, *non-dimensionally,* as any simultaneous state of matter, as a positively charged neuron. Which means, zero force or electrical potential difference has occurred.

DIMENSION 1 – You experience the restoral of isolated potential, as synchronizing or repeat whole number pattern, oscillating or repeating itself *non-dimensionally*.

DIMENSION 2 – You, as a derivative, experience the *restoral* of mutually interchangeable properties, as isolated potential, within any ascending whole integer, as length and width, which must ascend as one single fixed rate of conciliatory rotation, or 8 subatomic rotations, as sound waves. Disruption of force only repeatedly occurs when synchronized isolated potential, as oxygen, disturbs the rest point of oscillatory motion, exhibiting one continuous straight line of motion, as a positively charged positron.

DIMENSION 3 – You experience the *restoral* of isolated potential, as mutually inclusive properties. Which must isolate and restore a single solution auxiliary field, as a tri-auxiliary network, as repeat pattern, Meaning, oxygen must derive itself, within a fixed or pre-determined amount, in order to exhibit balanced pressure, within equal amounts of oscillatory motion. Which means a pre-determined amount of conciliatory rotation exists within **one coordinate**, as the mutually interchangeable properties of light, within a pre-established rate of sound, as a negatively charged neuron.

DIMENSION 4 – This is where you experience *duration*. Meaning, there is no division of mutually exclusive properties, but wave pattern exists simultaneously within one *ascending* harmonized standing wave pattern. Which simultaneously repeats oscillatory motion, as an isolated or single fixed rate of motion, as duration of sequential order, as one continuous straight line of motion.

DIMENSION 5 – This is where you exhibit force, within different amounts of conciliatory rotation, at the rest point of oscillatory motion, by decreasing the apex within fixed percentages that rotate a positively charged neutron, isometrically. Which you then isolate or collapse as the fourth coordinate, for you to perceive as dimensional particle reality or **mass.** In other words, oxygen does not disturb the rest point of oscillatory motion. Which means force has not exhibited equal amounts of pressure, within conciliatory rotation.

DIMENSION 6 – This allows you as a derivative to isolate between oscillating sequential order, in any likely or unlikely *ascending* order. Meaning, through isolating a single fixed rate of charge, exhibiting one continuous line of motion as force, at once, an ascending frequency either restores the mutually inclusive properties of light within sound repeating, isometrically, as other alternate wavelengths isolating the geometry of one. Or a single fixed rate of charge repeating *rationalizes* whole integers ascending in sequential order, as mass.

DIMENSION 7 – Allows any beginning or ending to your repeating oscillation, as one continuous line of motion distorts lost wavelength, which gives you additional fixed wavelengths, as restorative properties, as mass. But only if the fourth coordinate exhibits a fixed amount of isolated pressure, within conciliatory rotation, to release a single isotope.

DIMENSION 8 – Allows for all possible ascending and descending sequential order, as the restoral of lost wavelength. Meaning, force exhibits equal pressure within repeat volume of space, in order to ascend or descend as a positive integer, as the fundamental isolated properties of mass isolated as a positively charged positron. Meaning, at the subatomic level, oxygen restores 100% of itself isometrically, which must isolate potential as one positively charged positron.

DIMENSION 9 – Allows for the isolation of 8 fixed rotations, isolating or restoring all previous lost wavelength, at the rest point or apex of oscillatory motion, within any single isotope, as a positively charged neutron.

DIMENSION 10 – We are back to a single point in the system, where all imaginable possibilities exist, within 8 rotations of conciliatory rotation, restoring *one* point, as one continuous line of oscillatory motion, with no disruption of force. In other words, 8 subatomic rotations exhibit equal amounts of force within pressure, which means *two* isotopes are retained at the rest point or apex of oscillatory motion, as one positively charged positron.

The **tenth dimension** is critical for us to understand and master conceptually.

Anything and everything that is mutually interchangeable exists within a single point of this reality, as non-inclusive properties, but only within the restoration of oxygen, as the tenth dimension. And I/you/we only *exist* in that potentiality, within the **geometry of one** *repeating* itself, isometrically.

As a single fixed rate of motion, exhibiting isolated potential as a single solution auxiliary network, feeding into itself as a positively charged *neuron*. Which can isolate a pre-determined or fixed amount of motion, at the rest point of oscillatory motion, restoring itself isometrically as oxygen, in order to fundamentally *perceive* myself/yourself/ourselves, as separate.

In other words, within one isolated point, as the tenth dimension repeating the whole oscillation, as dimension *zero*, which is nothing *or* everything and everyone **non-dimensionally as oxygen,** the wave pattern fundamentally restores "itself" as mass, repeatedly or isometrically.

Simply put, as one whole repeat vibration, exhibits equal amounts of oscillatory motion, as repeat volume of space, the geometry of *one* has **zero** electrical potential difference within less of itself, as restorative properties. As one whole fixed rate of conciliatory rotation, at the rest point, as mutually interchangeable properties as **dimension 10.** Which must derive itself *non-dimensionally*, within **dimension zero**.

This must mean the magnetic properties of light and sound are interchangeable, but only at a synchronized rate of restoral, as mutually interchangeable properties synchronizing lost wavelength at the rate of speed within sound, as a single fixed or pre-determined rate of 8 subatomic rotations.

Saying the same thing differently. Dimension zero is nothing *dimensionally,* as mass, within one point. Or everything and everyone *non-dimensionally*, which simultaneously exhibits itself as lost or descending wavelength restoring equal amounts of isolated pressure, within repeat volume of space, as mutually interchangeable properties.

It is the *dispersion* of a single fixed rate of charge as amplitude, which either *allows,* or cancels, a single solution auxiliary field, as repeat pattern. Meaning, in order for the wave to oscillate freely to distort or attract wavelength, at the rest point of oscillatory motion, the disruption of motion must allow a *simultaneous* or "repeat" **positive** *feedback* loop, by not decreasing the apex of conciliatory rotation, which releases the pressure.

I think we all know shorter wavelength isolates particles as solid matter, which must extend in wavelength to attract gaseous substances. Through isolating a single fixed rate of charge, isometrically, the quantum wave pattern can isolate any fixed amount of restorative properties.

Which mutually extends the boundary of time and space, as mutually inclusive properties, as an *open* system. You are a positive feedback loop, rerecording a single fixed rate of isolated charge, as a repeat unified field, as pattern, as the whole field synchronizes the restoral of oscillatory motion, but only at the rest point of conciliatory rotation, does the apex allow fixed percentages restoring lost wavelength, as whole resonance, as the restoration of resonant frequency, as mass.

BECOMING A QUANTUM SUPER PROCESSOR

When all of our brain waves exhibit balanced pressure, as one continuous straight line of undisturbed motion, 8 subatomic rotations of oxygen maintain equal amounts of oscillatory motion, within the geometry of one repeating, as a single fixed rate of charge. Which means as a single solution tri-auxiliary network, we isolate gamma waves, as one **beta plane** pre-determining a single fixed rate of isolated potential, for future input of all preceding operations.

Meaning, as a repeat ionized state retaining oscillatory motion, at the rest point or apex of conciliatory rotation, balanced pressure is simultaneously repeated as a single cell of whole harmonious frequency, restoring one single amplitude, as an ascending harmonized standing wave pattern. In other words, **pitch** *becomes* isolated as a positively charged positron retaining the geometry of one, isometrically. Which means our theta waves keep us emotionally connected and aware of our oneness or being everyone and everything, delta waves keep us in the present, and our beta waves become amplified as an ascending harmonized standing wave pattern, repeating a single solution auxiliary field, as a tri-auxiliary network, feeding itself isometrically.

We must learn to isolate our brain wave state, in order to unilaterally *distribute* a single fixed rate of charge, by becoming aware of how we feel. We then can magnetically restore dimensionally particle reality, as mass, within a repeat rubric that resonates with us, as mutually interchangeable properties, over random disorder or chaos.

It is important to note when we have a stressed perspective, we also release chemicals that isolate us even more, as they increase the frequency of our beta waves. Those chemicals are what keep us safe when we are in real physical danger. Nature collapses all of our focus into the information within beta waves, so that we can process the next "logical" step to take that will keep us physically safe and out of danger.

Our brain cannot separate what we *perceive* as emotionally dangerous or what is actually really physically dangerous to us. If we feel emotionally unsafe and it stresses us out? We can then be in a "beta wave box" perceiving this reality from a distorted perspective, diminishing our integrated wave power, at the rest point of oscillatory motion, as one continuous line of motion, by decreasing the apex of conciliatory rotation.

We each are a unique resonant frequency, as an open living system, within one wave exhibiting the properties of oscillatory motion, which allows us to have a fundamentally fixed or distorted experience of reality. As a derivative, or rate of change, we are **whole** as 100% of one continuous line of motion, which *infinitely* restores itself within the geometry of **one,** until you as a derivative, cannot isolate mutually interchangeable properties, *isometrically.*

We are a whole repeating number pattern establishing any possible ascending order, but only within an isolated or single fixed rate of sound, rasterizing or converting the speed of light, as the restoral of our own unique resonant frequency. Which we must derive a rate of change to restore, as a rate of fixed percentages, within a single solution, tri-auxiliary network, as repeat pattern.

Meaning, if we do not retain balanced pressure within the wave pattern, we cannot isolate thermal energy, to retain a positive feedback loop. Saying it differently, you do not want to lose pressure from the system, by not retaining a single fixed rate of charge, at the rest point or apex of oscillatory motion, which pre-determines a fixed rate of conciliatory rotation, as an isolated metric of 8 subatomic rotations of oxygen.

Simply put, you will not retain *perpendicular* drag, which allows balance pressure by retaining 100% of thrust. Which means light *can* restore itself as mutually interchangeable properties of sound, within one repeating itself, isometrically.

In other words, a single solution auxiliary field cannot subdivide itself without restoring mutually interchangeable properties, within a mutually extended boundary of time and space.

To apply this level of understanding to our experience of daily life, you and I cannot be having the same experience of oscillating information, if one is not isolating itself isometrically. However, as a derivative or rate of change, either one of us can exclude each other, at the loss of our own utility.

It is critical for us to master that at the subatomic reading or level, we each co-exist dependently and interdependently, as a pre-determined amount of oscillatory motion, at the rest point or apex of conciliatory rotation. Meaning, the level of oxygen in the system is critical for us all to retain within our own experience, and we cannot negate the responsibility of maintaining the overall health of the system, as a whole.

To put it quite simply, if we do not retain lift, as one continuous line of motion displacing itself at the subatomic level or reading of 8 subatomic rotations, at the rest point or apex of conciliatory rotation, we deplete the system of subatomic energy.

MITIGATING LOST INFINITE POTENTIAL

As isolated potential repeating itself within the geometry of one, isometrically, to *simultaneously* restore lost or *divided* wavelength, we are each a positive feedback loop restoring mutually interchangeable properties. But only in the present moment with zero electrical potential difference. Meaning, as whole integers *replace* lost amounts of oscillatory motion **concurrently,** at the rest point of conciliatory rotation, we must equally ascend at the **apex** of oscillatory motion, as one single amplitude.

You want to have your awareness present, feeling aligned to your purpose because it resonates with you, because of how it makes you feel, in order to *ascend as* mutually interchangeable properties, as dimensional particle reality, as mass. It is a **mindset** to become aware of a single fixed rate of isolated potential subdividing itself infinitely, as any expression of itself fundamentally restoring dimensional particle reality, as mass

It requires "**organizing**" your awareness to develop skills that allow you to process isolated potential spontaneously organizing within the whole vibration, at a single fixed rate of speed, as simultaneous chemical and kinetic reactions combusting at fixed percentages. You must be at the same *velocity* to rasterize or convert resonant frequency, as your ascending simultaneous frequency responds to your own resonance.

Meaning, any possibility that exists within isolated potential, at a single fixed rate of speed, must repeat its own oscillation to isolate a derivative of it, as a positive feedback loop for future input of all preceding operations. Which means dimensional particle reality, as mass, resonates within sound restoring the mutually interchangeable properties of lost wavelength, at the speed of light. As a complete rotation exhibiting force, as a single fixed rate of charge, in order to isolate balanced pressure.

We are so often locked into rigid ways of being or doing based on experiencing 5% *less* than 100% of lost wavelength, as a whole single solution repeat vibration. This keeps us in a beta amplitude where we mitigate the charge, by descending in amplitude, as multiple vertices restoring lost resonant frequency. Which decreases the pressure, at the rest point or oscillatory motion, at the apex of conciliatory rotation. This is *why* neutrinos do not oscillate at full rotation or oscillation.

There is 100% of ascending wave pattern, as a single solution, tri-auxiliary network retaining oscillatory motion, as force within balanced pressure restoring lost wavelength that we do not view or "see" as reality, if we mitigate less than 100% of a single fixed rate of charge, repeatedly. Meaning, by only rasterizing or converting 5% as normal matter, we do not have the power to restore lost wavelength, by mitigating the charge. Which means less amplitude is repeating within less of itself, within a single fixed rate of percentages restoring an optical illusion.

Which challenges our ability to isolate the restoral of infinite potential, as a single fixed rate of charge simultaneously repeating itself as a positive feedback loop for future input of all preceding operations.

Put quite simply, isolated potential exhibits force within *balanced* pressure randomly accruing balance of charge, within any positive chain reaction retaining simultaneity, as an ascending harmonized standing wave pattern, repeating itself, as a positively charged positron. But in order to *visually* isolate the restoral rate of the speed of light oscillating within the rate of sound, we need to synchronize **pitch**, within volume of sound repeating.

AN EMPOWERED CONNECTION LIMITS POLARIZATION

We each exhibit force as undisturbed oscillatory motion, at the rest point of conciliatory rotation, which we distort as dimensional particle reality emotionally, conceptually and perceptually. Through those experiences of reality, we each have the ability to remain **grounded,** in order to isolate or rasterize dimensional particle reality in the most empowered way. In other words, nobody else can release our electrons.

To become a fully charged and harmonized *complete* circuit of information, as future input for all preceding operations, at the rest point of oscillatory motion without disrupting one continuous line of motion, at the apex of conciliatory rotation, is the only condition limiting synchronized wave pattern. Meaning, as isolated potential spontaneously accrues at 90° angles, we either mitigate or complete the balance of charge. Meaning, we propagate the transmission of one, isometrically, within itself repeatedly, or we descend into less of one, repeatedly.

Becoming aware of our state of being when we experience reality allows us to "supercharge" our rate of receptivity. To do this, we want to readily become aware of our emotional state, our conceptual state and our perceptual state.

In other words, we want to notice *how* we feel, and if we are *allowing* types of conditions to limit our ability to construct reality conceptually and learn to become *aware* of any limiting perceptions of others and our own unique sense of self. This allows us to retain balance of charge through the release of electrons, within an open and harmonized *ascending* rate of flow of information as energy.

It is when we feel powerless that we become angry or resentful towards each other as derivatives, which disallows our empowered state as the whole vibration, and our ability to become a harmonized standing wave as repeat pattern. In other words, we cannot mutually ascend as interchangeable properties, replacing "lost" or descended frequency, if we choose to insulate our electrons based on how we *feel*.

In other words, when our reception or awareness of this reality is **divided,** over *harmonized,* we cause deconstructive wave interference. Meaning, we cause a gap in volts or electrical potential difference within the rate of flow as energy, which mitigates the charge and distorts the pattern. Which means we cannot isolate mutually interchangeable properties, as one single amplitude, as a simultaneous feedback loop for future input of all preceding operations. We must distort the wavelength within a single amplitude, at the rest point of oscillatory motion, without disturbing the apex of conciliatory rotation, in order to repeat an ascending harmonized standing wave pattern, as mutually interchangeable properties, if we do not want to decrease the pressure in the system.

Saying it a different way, if we become an *incomplete* circuit as repeat information, we cause a negative feedback or distortion, which means the rubric must descend into fixed percentages, within less of itself, proportionately or isometrically. Otherwise, whole integers cannot replace ascending harmonized wave pattern to retain balanced pressure within the system, to avoid retracting within a single point in the system. This means force is used inefficiently to maintain velocity within an *expanded* volume of space.

Retaining a repeat harmonized or unified field pattern within our human experience of oscillating reality is challenging. That is why if we ground or receive at least 50% as the whole vibration, the other half retains **charge**, *through* the experience of itself as complete. Which allows electrons to freely relay as whole integers repeating in a positive feedback loop.

DEMANDING YOUR ONE WAY OF REALITY

We so often insist that reality be experienced in a way that allows **us** to *feel* right. And what I mean by "right" is for me to feel in balance as the *whole* vibration. Because if I feel you are wrong, I cannot become harmonized within the whole vibration *with* you. My judgment **disallows** my balanced state or whole single solution auxiliary field, as a *positive* loop feeding back into all future preceding solutions, if I feel you are not a combined possibility. Based on the **condition** I *demand* you to experience, in order for me to feel harmonized, as an *ascending* harmonized standing wave, as a repeat pattern or one repeating itself, isometrically.

"Right" can take on many different forms, whether it is the "right" ideology, the "right" behavior or the "right" choice. If you disallow your resonant vibration to others or groups of people based on your need to feel right, you cannot retain an empowered rate of flow as subatomic energy. Meaning, you cannot retain balance of charge within negative feedback distorting lost amplitude, within less of itself, repeating within a fixed rate of proportions. In other words, you mitigate charge, which decreases the pressure, by reducing the amount of subatomic rotations of oxygen, at the rest point of conciliatory rotation.

You then cannot experience rationalized potential, as repeat whole number patterns retaining fixed percentages of the infinite expression of length, width and height, until one as a positive integer can no longer repeat itself isometrically, as a single fixed rate of proportions. In order to do this, you must retain an *ascending* harmonized wave pattern, as a single solution auxiliary field, as a tri-auxiliary network isolating repeat potential, as rations of whole numbers resonate within your unique expression of dimensional particle reality, as mass.

Quantum reality can feel abstract, within a disempowered reality experiencing 5% *less,* than 100% of your resonant feedback existing within a single amplitude, as one simultaneous **positive feedback loop,** as future input for all preceding operations. Meaning, a simultaneous reality does exist that allows you to co-exist, each feeling you are experiencing reality in a way that works for you, *without* feeling limited. Those ascending states of matter exist within resonant frequency, as an infinite point within the tenth dimension, where there is *no* deconstructive wave interference as the whole vibration, relative to dimension zero, or nothing *non-dimensionally.*

Dimension zero again is nothing **dimensionally** *within a single point*, which is **everything and everyone** *dimensionally* at the rest point or apex of conciliatory motion. Which allows anything imaginable, under any condition a Universe can be created under, through restoring lost or distorted wavelength, as your own unique resonant frequency. You just have to allow a single fixed rate of isolated potential, by releasing your electrons and ascending, as a repeat harmonized standing wave pattern.

In other words, we are all exhibiting raw potential, out of everything imaginable, with no electrical potential difference to ascending frequency restoring magnetic properties. Saying it a different way, we are truly a dimensional particle reality, out of *anything* **imaginable,** using our resonant frequency as simultaneous experience, within a repeating or ascending *harmonized* wave pattern, repeating 100% of the information.

The difference is everything. Meaning, we either cause a gap in volts and repeat within less of ourselves *proportionately,* or we exhibit a unified force, which does not exceed the rate of potential that would allow one, as mutually interchangeable properties, to supersede any and all combined possibilities.

RETAINING POWER THROUGH THE RELEASE OF ELECTRONS

Ultimately, it is challenging to remain in an open state at all times in this reality. We are human. We have feelings. And sometimes we make choices or have feelings that challenge our ability to retain an open unified field or auxiliary pattern, as any combined solution, at a fixed rate of speed. We can then mitigate or isolate a massive gap in volts, as electrical potential difference reoccurring within negative feedback, as the *repeat* of lost wavelength not being fundamentally restored.

The power of maintaining compassion and empathy for choices others make as an outcome of their own polarization or loss of resonant frequency, is that *you remain* **consistently open.**

And if you enter into a state of polarizing your electromagnetic field, or distorting a single ascending amplitude? You have *retained* a tri-auxiliary network, oscillating at a single fixed rate of motion, that does not disrupt the rest point of oscillatory motion, by disturbing one continuous supply of oxygen, at the apex of conciliatory rotation.

Which is any combined solution, at a fixed rate of speed, oscillating within a repeat pattern of ascending whole integers, repeating as an *ascending* harmonized standing wave pattern. But only **if** you retain and release a positively charged positron, as your own resonant frequency.

In other words, all the information is amplified, as one complete circuit of information, over deconstructed as incomplete, within any simultaneous loop of whole integers rasterizing or converting dimensional particle reality as mass, faster than force exhibiting balanced pressure within itself, isometrically. Simply put, you cannot destroy resonant frequency, and isolate infinite potential, as the completion of one restoring itself, as an isolating metric.

Rasterizing or converting distorted or descending frequency is *mitigating* or "lessening" the rate of charge. Meaning, there is no accrual of resonant frequency lost within the system, as repeat pattern isolating infinite potential. Infinite potential again, is the repeat or the recycling of a single fixed rate of charge, within any whole or repeat oscillation, at the rest point of oscillatory motion, within an isolated positive feedback loop, which does not decrease the pressure in the system. Meaning, the disruption of one continuous line of motion has been disturbed at the apex of conciliatory rotation, as the *withdrawal* of oxygen, as **raw potential.**

However, if the loop exhibits fixed amounts of equal pressure, by releasing and retaining a single a positively charged positron, the law of motion has not been violated by not exhibiting force that supersedes balanced pressure, as an isolated repeating rubric synchronizing the restoral of resonant frequency. Which means, one continuous straight line of motion, at the rest point of oscillatory motion, must isolate 8 fixed subatomic rotations as energy, at the apex of conciliatory rotation.

In other words, 100% of one does not isolate a single fixed rate of derived charge, if a negative integer magnetizes the restoral of dimensional particle reality, as mass.

Isolating charge is a repeat mechanistic process of retrieval of a single repeating wave amplitude. In theory, this single wave amplitude can restore any compound integer, by isolating a repeat unified matrix as a single solution auxiliary field, repeating itself as pattern. But only within a pre-determined rate of fixed percentages that reduce a positively charged positron, into a negatively charged neuron, as a positive feedback loop for future input for all preceding solutions. Which rationalizes any sequence of mutually exclusive properties, as positive whole integers simultaneously unite and divide lost wavelength.

TEN WAYS TO RESTORE YOUR RESONANT FREQUENCY

We must master a mindset to *retain* mutually inclusive properties, without mitigating charge repeatedly by disturbing the rest point of oscillatory motion, at the apex of conciliatory rotation. Saying it more clearly, we do not want to repeat less of whole charge, as a fixed rotation of conciliatory motion, which distorts or descends frequency within **unequal** *amounts* of oscillatory motion. Motion is then lost from the system, which *decreases* the apex of conciliatory rotation.

If we cannot complete the balance of charge to restore our lost distorted wavelength, as a repeat pattern synchronizing simultaneous combustion cycles as non-dimensional states of particle reality ascending in order, we disturb the rest point of oscillatory motion, and we disrupt one continuous straight line of information. Simply put, we want to replace a single fixed rate of charge, without destroying 100% of it, by mitigating it.

1. STAY PRESENT

There is no time. Time is an illusion. One whole repeat
combustion cycle, as a full or complete rotation of
conciliatory rotation, rationalizes whole integers. Which
means sound waves are interchangeable within the speed
of light traveling equidistant of each other, but only in the
present moment. If your awareness is consistently in the
future with fear, or in the past with judgment, it is very
difficult to retain oscillatory motion, at the rest point
or apex of conciliatory rotation.

Which means your resonant frequency is descending as lost
wavelength, distorting oscillatory motion in unequal amounts
and decreasing pressure. In other words, you cannot hear or
see lost potential infinitely rasterizing or converting 100% of
lost wavelength, as one whole repeat oscillation, as a single
solution auxiliary field, repeating the **geometry of one,**
isometrically, as a tri-auxiliary network.

2. JUDGMENT IS LESS THAN PRECEDING AMPLITUDE

Clean up your perception by perceiving others as a source
of infinite potential. Check in regularly during your
interactions and become aware of *your* perception of others.
Are you limiting others, as a rate of mutual exchange, based
on age, race, sex, education, bank account, social status
or professional experience? This translates into
deconstructive wave interference and you mitigate charge.

Also become aware of your perception of you.
Do you feel limited in any way?

The key to becoming *discrete* potential energy or quantum
is to not feel limited in any way, and to know you are a
source of infinite possibilities, as one continuous straight line
of motion, but only as a single solution auxiliary field,
repeating as pattern.

In other words, within one whole repeat oscillation spontaneously combusting, at a single fixed rate of isolated potential, as a positive feedback for future input of all preceding operations, you are any combined possibility repeating within you, as a single fixed rate of charge. However, you must synchronize *force* within balanced pressure, in order to charge the release of a single isotope. Otherwise, a positively charged positron will not exhibit one continuous straight line of motion, with no stopgap or disruption of motion, as one simultaneously restoring itself, as one *complete* or a negatively charged neuron.

Only when one continuous straight line of motion is disturbed at the rest point of oscillatory motion, will the apex of conciliatory decrease. Which means the boundary of time and space has *exclusively* extended beyond mutually interchangeable properties. However, in order to experience ourselves as different or *separate* and still retain a positively charged positron, we must ascend within positive integers, as a repeat harmonized standing wave pattern. This is only possible if we perceive reality as an open system, *consistently*, through the release of electrons.

3. ORGANIZE YOUR AWARENESS

To "organize" means to feel and process oscillatory motion *organizing* oscillating reality, as a single fixed rate of charge simultaneously repeating itself. Meaning, a synchronized electromagnetic field allows you to isolate potential becoming possible, as helium rises, when oxygen, as a net worth of zero, *subdivides itself* **8 times,** at the rest point of oscillatory motion, at the apex of conciliatory rotation. Which means *non-dimensional* oscillating frequencies resonating within your positive feedback or whole oscillation, *restore* your resonant frequency, as mass.

You feel oscillatory motion in your gut, then you receive it emotionally in your heart, and then you break it down further with your awareness, by using your brain.

If you start in your head, you are already in the probabilities, because you feel separate in your head, based on your *visual* experience of *preceding* reality.

That is your quantum power. Organize it to use it.

As challenging as it is to realize, what we see is an illusion of a pre-determined outcome, based on a single fixed rate of percentages, we must realize this is the output of less than 100% of mutually interchangeable properties being used as input for all future preceding operations. In other words, we have *inverted* lost wavelength. Thinking of it at the subatomic level, one cannot divide itself within 5% of lost wavelength, 8 times, in order to isolate a single fixed rate of oscillatory motion, at the rest point or apex of conciliatory rotation.

Regardless of what you *see*, you have the power to restore your own resonant frequency and retain oscillatory motion, at the rest point or apex of conciliatory rotation, to restore resonant frequency as sound waves. Which do not mitigate charge, repeatedly.

4. ATTACHMENT TO PROBABLE OUTCOMES

A big shift in awareness to becoming quantum is to let go of attachment. Any combined possibility that *resonates* can come from anywhere.

However, if you are attached to an outcome coming from a particular source which is not possible *within* a simultaneous positive feedback loop **synchronizing all combined possibilities,** as future input for all preceding operations, as mutually interchangeable properties, you

cause *deconstructive* wave interference.
Probable outcomes cause "distortion" and shorten the
wavelength. Which mean you cause *negative* feedback.

You can overcome a distorted feeling of separateness,
by isolating distorted wavelength as a positive feedback loop
for future input of all preceding operations. In other words,
you mutually extend the boundary of time and space by
becoming any combined possibility. Which does not exhibit
or *decrease* isolated pressure, or equal amounts of
oscillatory motion, within your own distorted reality. Meaning,
you do not *disturb* one continuous straight line of oscillatory
motion, at the rest point, which then can subdivide itself 8
times, *as a net worth of zero*, or oxygen. This means again,
there is no disruption or distortion of mutually
interchangeable properties.

Let go of attachment. Become quantum.

5. BECOMING YOUR UNIQUE SELF

There has never been anyone like you or will be anyone
like you. You are unique for a reason that exists now.
You, as who you are, has never existed, before now.

The question is, why are you, who you are?
What makes it possible to be you? What does the whole
oscillation what to become through you? What does the
whole vibration what to experience and bring forth, that only
you can become, for it to experience?

We so often feel limited because of our *perception* of what
is, right now. When we participate in mastering the
awareness of combined possibilities, we want to actively
participate in showing up as our unique self, in order to
derive 100% of lost wavelength, at the rest point of
oscillatory motion, to fundamentally restore conciliatory
rotation, at the apex.

When we have a deep sense of our own uniqueness,
we have a deep sense of valuing who we are. It's when we
realize all the possibilities are possible for us to become
more of who we are, that we fearlessly become who we are.

Why would we want to be anybody else as resonant
frequency?

6. STAY IN THE POSSIBILITIES

If I perceive you as less than restored potential, through you,
I *become* **less** than 100% of my resonant frequency. I will
not hear, understand, process or assimilate more than what
I perceive is possible, through you. In other words, in my
relationship to you, I will only stay in the probabilities
or the fixed information that I know.

In other words, I limit myself as a single fixed rate of charge
repeating or indexing a *whole* single solution auxiliary field,
as repeat pattern, mutually ascending as a single amplitude,
when I *perceive* others as less than a source of *ascending*
potential. Meaning, as a derivative or rate of change, my
force decreases within balanced pressure. Which means my
whole repeating index of heat oscillating as rhythm, as a
derivative of one or single fixed rate of charge, cannot
subdivide itself 8 times, if mutually interchangeable
properties are not non-exclusive properties. In other words,
any ascending whole integer must *replace* descending lost
wavelength, in order for a positively charged positron
to remain stable.

We so often "exclude" based on what keeps us safe or
feeling "right" when at the subatomic level, it is all one single
fixed rate of charge, deriving any singular possibility,
within simultaneous non-dimensional states of combined
possibilities. The key to becoming quantum as discrete
energy is to isolate the combined possibility resonating

with you, in order to mutually ascend, as a *stable* mutually interchangeable property.

7. SERVICE TO THE WHOLE

When you choose to become more of who you are, in service to the whole, it becomes easier to oscillate as a positive feedback loop. You then *amplify or "repeat" lost wavelength as oscillatory motion,* as you feel one with conciliatory motion.

In other words, you become **equivalent** as a single amplitude retaining 100% of spontaneous kinetic and chemical reactions randomly accruing. Which allows you to rasterize or convert any combined possibility, as a single fixed rate of charge. In other words, you do not cause negative feedback, at the rest point of oscillatory motion, by relaying within whole integers, which *retains* balanced pressure as your unique reality. You are a repetition of one whole cosmic frequency, isolating rhythm as repeat pattern, to fundamentally restore itself, as any combined possibility.

In other words, you *avoid* distortion or negative feedback as the *disruption* of oscillatory motion, by not exhibiting or converting the chemical compounds of descending lost wavelength. Which does not isolate repeat metrics, as a single fixed rate of charge, as a single solution auxiliary network for future input of all preceding operations. You must retain one single positively charged positron, at the rest point of conciliatory motion, as balanced pressure. Put simply, sound cannot travel through fractions of itself repeating, without distorting the *frequency* or **whole resonance.**

If you feel separate, you mitigate charge. You then isolate a **fraction** of whole integers relaying within distorted or shortened wavelength, which disturbs the rest point of oscillatory motion and decreases the apex. Which means you can no longer subdivide one, 8 times, as a net worth of zero, which *releases* helium and **retains** hydrogen.

Put simply, this is *why* sound travels through water in this reality, but not space.

8. TUNE IN

We have the ability to hear and feel oscillatory motion through resonant frequency, as sound resonating within equal amounts of oscillatory motion, or balanced pressure within force repeating itself, isometrically. But only if the rest point or apex of conciliatory rotation subdivides itself 8 times, as volume of space repeating isometrically, as *zero electrical potential difference or opposing amplitude,* as **undisturbed conciliatory motion.**

In other words, one whole repeat oscillation, as 100% of a single fixed amount of conciliatory rotation, is preceding input for all future operations. Which means an *ascending* harmonized standing wave pattern, *instantly* restores isolated potential, as mutually interchangeable properties fundamentally *restore* dimensional particle reality, as mass.

What this means is what is becoming possible is pre-determined according to action taken. To stay up to date, check in regularly by listening to your inner voice. Trust it. It has the latest information on what you need to do to become aligned to what is becoming possible, which will allow you to experience you in the most empowered way, in a quantum reality where anything can happen.

We are powerful creators, but first we must organize mutually interchangeable sequences of reality, in order to be able to rasterize or convert those combined possibilities, within 100% of mutually interchangeable properties.

9. RETAIN CONNECTION IN ALL YOUR RELATIONSHIPS

Normalize choices and values in your daily life, so you *retain* oscillatory motion, at the rest point or apex of conciliatory rotation as resonant frequency *through* your relationships, by norming your social behavior.

Value feelings, both your own and others. This keeps you emotionally aware and integrated, over isolated, which retains your harmonic ratios *through* maintaining empathy in your relationships. In other words, take care to be kind to yourself, and others, to maintain an open system, as one subdividing itself 8 times, through your relationships. Which means, it must subdivide itself 9 times, within mutually interchangeable properties, in order to not cause opposing amplitude.

You do not want to mitigate the charge of a positively charged positron and become a fraction of resonant frequency, repeating itself isometrically. You want distorted wavelength to oscillate freely, within fixed percentages of itself, as mutually interchangeable properties.

Listen to understand. This keeps you focused in the present moment, aware and exchanging whole frequency within a relationship.

Be curious. This keeps you free from assumptions, by being in a state of curiosity, which allows for the awareness or understanding which retains oscillatory motion at the rest point or apex of conciliatory rotation.

Always assume best intentions. This keeps you free from judgment and negativity and avoids mitigation of charge, which retains balanced pressure within the oscillation.

Always be willing to learn from any choice, made by you or others. This allows you to have unconditional acceptance of yourself and others. You cannot retain a positively charged positron, as an open system repeating itself isometrically, in a state of *conditional* living.

10. ALLOW QUANTUM REALITY TO COME TO YOU

Remember isolated potential will only oscillate or repeat a single fixed rate of charge, when *one* subdivides itself multi-dimensionally. In order to exhibit simultaneous force *within* isolated pressure, as any combined possibility resonating within your quantum field, as repeat pattern. Which then *restores* dimensional particle reality, as mass.

However, information being observed out of *preceding* wave pattern, must retain a positively charged positron, or a single fixed rate of charge does not magnetize itself, by repeating itself isometrically.

We must *retain* a positively charged positron, in order to *resonate* within mutually interchangeable properties, *as* **everything imaginable,** in order to be able to subdivide 100% of one, 8 times, as any positive integer, to retain force within balanced pressure.

Meaning, when force exhibits balanced pressure, by instantly restoring the *non-dimensional* preceding reaction, within whole numbers repeating a single solution auxiliary field, as repeat pattern, **equilibrium** is retained. Meaning any mutually interchangeable property ascends, as a restorative compound integer.

Which means, within an isolated or fixed rotation synchronizing repeat combustion cycles, whole integers will relay mutually interchangeable properties, as the geometry

of one, isometrically repeating volume of sound, as a positive feedback loop, which restores conciliatory motion.

All you have to *do* is stay repeatedly aligned to harmonic values at the rest point of oscillatory motion, which synchronize non-dimensional states of matter. Which would mean one repeat or *ascending* harmonized standing wave pattern oscillates or subdivides itself **9 times**, through the release of a single isotope, simultaneously replacing itself as a positively charged positron, subdividing itself 8 times, as oxygen. Which allows you to become more of who you are in a balanced and harmonious way.

To put it simply, "one" must refrain from exhibiting oscillatory motion, to exhibit mutually interchangeable properties, being subdivided 8 times, at the rest point of conciliatory rotation. This is why a positive feedback loop, *restores* the apex.

We have been focusing on our individual experience of quantum reality, in order to theoretically understand our experience of a dimensional particle reality as *infinite* potential. Now we must take a much deeper look into the system, as a whole or two-prong system, exhibiting equal amounts of pressure within volume of space, as a single solution auxiliary field, repeat as pattern.

PART TWO

QUANTUM REALITY IS A TWO-PRONG SYSTEM

4

ALL ENERGY IS RELATIVE TO 100% OF ONE, IN A QUANTUM UNIVERSE AS REPEAT PATTERN

THE BUILDING BLOCKS OF PARTICLE REALITY

A single solution must precede itself, as a positive feedback for all future operations, in order to close the gap of utility, as mutually exclusive properties exhibiting unequal amounts of one continuous straight line of motion.

This means a positively charged positron must exhibit itself *twice* or **simultaneously**, as a mutually extended boundary of time and space, in order to replace a negative integer, as an ascending harmonized wave pattern simultaneously *repeating* a positively charged positron. If not, a single fixed rate of charge does not isolate equal amounts of oscillatory motion, at the rest point or apex of conciliatory rotation. In order to subdivide itself as a net worth of zero, 8 times. Which releases both oxygen and a single positively charged positron exhibiting equal amounts of pressure by releasing helium and retaining hydrogen.

In other words, one continuous straight line of motion does not experience a gap of utility as differentials opposing itself, at the rest point or apex of conciliatory rotation. Which causes opposing amplitude or negative feedback and the wave pattern can no longer ascend within fixed percentages, as a single solution auxiliary field, repeating as pattern, in a two-prong system.

Moreover, light and sound are no longer mutually interchangeable properties, exhibiting equal amounts of distorted wavelength simultaneously restoring lost pressure. If oxygen is not complete as a substrate, sound will not magnetize at a rate of speed *equal* to the speed of light, retaining equal amounts of oscillatory motion at the rest point, without decreasing the apex of conciliatory rotation.

We must consider if one continuous straight line of motion is being *received* as a two-prong system restoring a magnetic shield, as a single fixed rate of charge *exclusive* of **entropy.** Meaning, we must consider is one continuous straight line of motion exhibiting the *loss* of oscillatory motion being *reflected* within mutually exclusive properties only, in this collective version of dimensional particle reality, as mass.

Meaning, is oxygen a substrate which breaks down subatomic energy, if equal amounts of oscillatory motion, at the rest point or apex of conciliatory rotation, do not simultaneously restore a repeating or ascending harmonized standing wave pattern as a single solution auxiliary field. Which restores itself within fixed percentages of one, within three dimensions of dimensional particle reality as mass oscillating within 100% of lost wavelength, feeding back into the system, in equal measure to be *refracted* within equal volume of space. Which logically tells us, must repeat until a single fixed rate of charge, *repeats.*

Going back to nature as an example, photosynthesis is the outcome of decomposition or decreased pressure, within repeat volume of space. Meaning, if oxygen had a net worth of 8 subatomic rotations at the rest point or apex of conciliatory rotation, as rations of fixed or equal amounts of one continuous straight line of motion, it would not need a substrate or enzyme to multiply its distorted or shortened frequency.

Saying it a different way, if the co-factor of *zero* as **-1** replaces itself as **+1**, then there would be zero entropy or waste exhibiting unequal amounts of balanced force. Meaning, equal amounts of one continuous straight line of motion would continuously charge an *ascending* positron.

This is why the substrate of **-1** causes a continuous radioactive pattern throughout the Universe, as force within *unequal* amounts of balanced pressure, which decreases the apex of conciliatory rotation. Which of course means, a single fixed rate of charge no longer isolates itself, as a positive feedback loop for future input of all operations, if the rest point of oscillatory motion has been disturbed.

In other words, magnetic properties must ascend as mutually interchangeable properties, as a single fixed rate of change, exhibiting one continuous straight line of motion.

CAUSALITY AS A SINGLE SOLUTION AUXILIARY FIELD

A single solution auxiliary field will not remove lost wavelength, if 100% of itself *restores* lost wavelength, as a positively charged positron. However, if 100% of itself does not repeat, lost wavelength is destroyed as mutually interchangeable properties exhibiting equal amounts of oscillatory motion, at the rest point or apex of conciliatory rotation.

Which means charge must *descend* within mutually exclusive properties, as a negatively charged positron, in order to retain a single fixed rate of charge, within fixed percentages by using the apex of conciliatory rotation to feed into itself, *proportionately.* As a single fixed rate of the speed of light within distorted frequency exhibiting equal amounts of motion, as the **x axis** *simultaneously* rotates to receive the **y coordinate.** Which means, one continuous straight line of motion no longer replaces a negative integer, if 100% of thrust is not used for future input of all preceding operations, to *retain* perpendicular drag.

One, as one continuous straight line of motion, must exhibit itself twice or *simultaneously*, in order to release and retain a positively charged positron.

To start, **A** as a positive integer, must coincide with **B,** as a negative integer. Together as *one* single fixed rate of the mutually interchangeable properties of light, exhibited as a fixed rate of sound, **A** and **B** must oscillate at the same rate of speed.

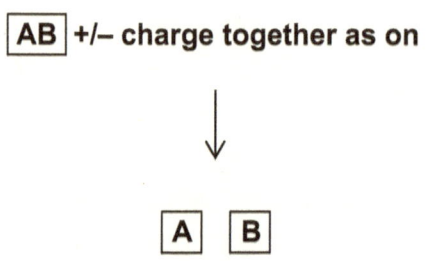

Charge is separated only when
mutually interchangeable properties
can no longer isolate a single
positively charged positron,
repeating 100% of one

Once charge is *separated,* there is zero potential difference as oscillatory motion. As a single solution, **A** will ascend in charge as the **x axis,** while B remains isolated as a single remaining positively charged positron. **A** will then mitigate itself, unless **B** retains the rest point or apex of conciliatory motion, as mutually interchangeable properties, as the **y axis.**

Unless the rest point of oscillatory motion, at the apex of conciliatory rotation is disturbed, which means one continuous straight line of motion has been disrupted. This means again, **A** or "one" can no longer subdivide itself 8 times, as oxygen releases a single positively charged positron. In other words, a single isotope must simultaneously be released, and positively restored, as mutually interchangeable properties, without subdividing less of itself, repeatedly.

If one is restored *as equal amounts of one continuous straight line of motion* in a single solution auxiliary field, **A** is then fundamentally able to repeat itself 8 times, exhibiting force concurrently as balanced pressure, as **B** isolates a single repeating charge. Putting it simply, **8** *subdivided* **exhibits** a negatively charged neutron, exemplifying force as one continuous straight line of motion, with no stopgap, or disruption of motion.

In other words, one atom must *simultaneously* split itself, *twice*. Saying it a different way, **8 ÷ 2 = 4 subatomic rotations,** but only at the rest point of oscillatory motion, with no disruption of charge, at the apex of conciliatory rotation.

There is no difference of charge as **4 ÷ 4** is inclusive of "one" infinitely repeating as a lost integer or isotope, as **100% of the volume of space** that retains one continuous straight line of motion, at the rest point of oscillatory motion, as a single fixed rate of 8 subatomic rotations. But only if force exhibits balanced pressure, or **A** oscillates within **B,** with no mitigation of charge.

In other words, we must isolate **"4"** as **zero** electrical potential difference, as the mutually inclusive properties of **(x)** and **(y)** synchronizing time and space, within any "one" point of a single amplitude of ascending whole integers. Which must exhibit force within balanced pressure, by oscillating within mutually inclusive properties, as an ascending harmonized standing wave pattern *restoring* a single solution auxiliary field, repeatedly.

In other words, **one** rests or isolates as itself, within itself, repeatedly at the ionic level of **ZERO**, repeating or restoring itself infinitely, exhibited as force within balanced pressure or equal amounts of the rest point of conciliatory rotation, but only at the apex. If the rest point has been disturbed by a **mixed** *rate* of fixed proportions, the pressure of the system decreases.

A simple flowchart will show you *how* fixed percentages repeat isolated oscillatory motion, as a single fixed rate of charge, infinitely *restoring* itself within mutually interchangeable properties. However, when force exhibits lost or distorted frequency as descended wavelength, simultaneously repeating itself, as a single fixed or pre-determined amount of conciliatory motion, light is reflected, over refracted. Which means *one* **is** not fundamentally restored, as 100% of lost wavelength, as preceding input for all future operations. It works both ways.

As a single fixed rate of charge or one subdividing itself 4 times, at the rest point or apex of conciliatory rotation, a positively charged positron releases helium, which pre-determines lost wavelength as the restoration of itself, infinitely. Until a positively charged positron restores itself *isometrically.*

A simple way of seeing it is tempo, or the speed at which the mutually interchangeable properties of light and sound exhibit force *unilaterally* within balanced pressure, as **ionic** *tempo.*

A quantum reality must allow for particle rotation, within subdividing itself as less of itself, within mutually exclusive properties, in order to allow for equal amounts of one continuous straight line of motion, at the rest point of oscillatory motion, without decreasing the apex of a fixed amount of subatomic rotations isolating mutually interchangeable properties.

In other words, a quantum wave pattern, as a single solution auxiliary field must **repeatedly** *allow* for any combined possibility within one *and* allow for any probabilistic combination. Which means 3 subdivides within itself proportionately, exclusive of one repeating, as a single fixed rate of charge simultaneously restoring itself, as a positive feedback loop.

THE LAW OF FIXED PERCENTAGES REDEFINED

One must exhibit itself 8 times as oxygen with a net worth of zero, for a single atom to split or subdivide force as the restorative properties of oxygen. In other words, one must allow for probable change, as it allows for the rasterization or conversion of itself within any combined possibility, as mutually interchangeable properties of itself, repeating infinitely. But only until one can no longer disturb or isolate one continuous straight line of motion, at the rest point or apex of conciliatory motion, as an infinite expression of itself.

$$4 \div 3 = 1.3333333$$

$$\downarrow$$

A positive single charge simultaneously repeats, and restores mutually interchangeable properties, within a single fixed rate of percentages coordinating X and Y as a single solution auxiliary field which pre-determines Z.

Consider .01 remaining as 100% of the rest point or apex of conciliatory motion.

$$(x)\ 33\%\infty$$

$$(y)\ 33\%\infty$$

Exhibits mutually interchangeable properties of light traveling equidistant within the rate of sound, as a pre-determined amount of one positively charged positron exhibiting a positive feedback loop for future input for all preceding operations.

(z) 33%∞

Think of it this way. Descended wavelength must exist within itself 4 times, as any mutually interchangeable property, and then subdivided by 3, to retain a single fixed rate of charge repeating. Which exhibits any mutually interchangeable properties, as one continuous line of motion, *repeating* itself infinitely, with no stopgap. *Unless* one continuous straight line of motion is disturbed at the rest point of oscillatory motion, at the apex of conciliatory rotation.

There will always be *one* remaining **proportionately**, within the rasterization or conversion of 3 simultaneously converting itself, as any mutually interchangeable property, within one point. This means "one" positively charged positron, exhibits a single fixed rate of charge.

But only at the rest point of conciliatory motion, will the mutually interchangeable properties of one, subdivide itself 8 times, without disturbing the rest point of oscillatory motion, at the apex of conciliatory rotation.

Which means two positively charged positrons, isolate synchronized combustion cycles exhibiting a single fixed rate of pre-determined force within balanced pressure, as mutually interchangeable properties. Which means **A** must then rasterize or convert **B**, as fixed percentages of one single amplitude restoring the geometry of one, isometrically, as future input for **all** preceding operations.

100

100% of one lost integer must exchange,
a single fixed rate of conciliatory motion,
as one rests within the tempo of one single,
or isolated fixed rate of speed,
as one whole repeat oscillation of oscillatory motion,
or fixed amount of pre-determined oscillatory rotation,
as 8 subdivided 4 times, repeatedly.

Which must pre-determine Z,
as the ascending wave cycle simultaneously repeats
or loops 99% of the electromagnetic properties
of "one" repeatedly, as preceding input
for all future operations.

What we may or may not realize, is that non-dimensional properties of one, exhibit one continuous straight line of motion. But only as a **single** *amp*, repeating itself *isometrically*, can we resonate as any combined possibility.

I think we have radically miscalculated the restorative potential of our quantum universe, by believing infinite is endless or "waste". Which then does not redistribute itself, as any combined possibility, within mutually interchangeable properties.

LOST WAVELENGTH EXHIBITING ITSELF AS A MODEL

Joules is not a repeat measurement of **force times distance**, as a single fixed rate of charge repeating a pre-determined amount of conciliatory rotation. The rate of energy within a model must be measured in **amps,** not *kilowatts.* We want to **isolate** how much of preceding fixed rotation is being used within an ascending wave matrix, as one continuous rate of sound isolates a fixed amount of percentages, exhibited as one continuous line of straight motion, with no stopgap.

Meaning as charge rests within itself, at the apex of conciliatory rotation, as a single fixed rate of repeating charge, as one whole amp of a single fixed rate or pre-determined amount of subatomic rotation, light will exhibit a gap in pressure as *descended lost wavelength* simultaneously feedbacks, as any mutually interchangeable property.

The rate of one continuous straight line of motion **must** flow through the *ascending* state, at the restoral ratio, which magnetically closes the gap. In other words, in order for the model to *recycle* the repeat volume of electromagnetic energy from its **descending** state, as future input for all preceding operations, the model must **subdivide** one whole or fixed amount of conciliatory rotation. In order to repeat itself within **one** ascending amplitude, as any mutually interchangeable property.

The whole state is any state of matter retaining isolated or magnetic properties, out of pure *raw* potential that magnetizes instantly with zero inertia. Meaning, pre-determined wavelength can oscillate or isolate anything, out of *nothing*, which is *everything*, fundamentally.

If we look at separating exhibiting properties from latent properties, **charge** *isolates* conciliatory rotation, as one single fixed rate of pre-determined proportions, as a single fixed rate of charge, repeating itself as oscillatory motion or one continuous straight line of still motion. Which must retain a magnetic shield, as an *ascending* harmonized standing wave repeating oscillatory motion, which does not *destroy* magnetic properties. Instead, one repeat of a single fixed rate of oscillatory motion *charges* a positively charged positron, as a negatively charged neutron.

The model will *divide* the dimensional system (10) in half which *separates* the raw potential flow by charge according to the law of inertia. Which states if motion is moving at a straight line, it will remain isolated unless exhibited upon by an outer force. Ten can be divided equally without reducing lift, which isolates or re-distributes a single fixed rate of charge.

The model then provides a single fixed rate of oscillatory motion restoring lost magnetic properties, which must close the gap of conciliatory rotation, or one subdividing itself 8 times, to release and retain a positively charged positron. In other words, sound is oscillating within the rate of the speed of light traveling *at a fixed rate or pre-determined amount of conciliatory rotation*, within equal measure of sound repeating.

A fundamental or restorative model must exhibit force as *one* single point within a single solution auxiliary field, as a repeat pattern, in a tri-auxiliary network isolating inertia by replacing conciliatory rotation. Only then can we presumably understand and see why any state of matter is fundamentally possible, by being mutually interchangeable, within geometry of one repeating itself, isometrically.

This is *why* our current standard model does not repeat itself as the geometry of one restoring itself, isometrically. We are taking one, as a substrate of itself, and dividing it evenly, without giving a thought to the infinite remainder. Which can be isolated as any combined possibility by fundamentally restoring itself, infinitely, as a pattern of compound integers, *ascending* at a pre-determined rate of fixed speed.

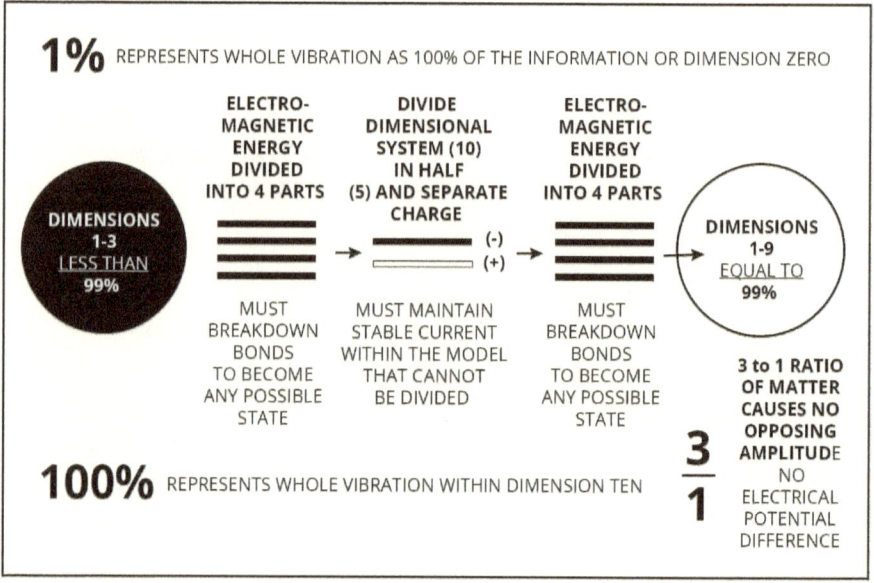

A QUADRATIC TRIANGLE ISOLATES DIFFERENTIALS

A single isotope does not repeat, unless charge exhibits mutually *inclusive* properties. But a positron will only fundamentally restore mass, if a positively charged positron can isolate a single fixed rate or pre-determined amount of conciliatory rotation, at the rest point of oscillatory motion, without decreasing the apex.

If we look at the base of a quadratic triangle, we only see mutually inclusive properties exhibited as **4 points** within any unilateral equation. A single solution auxiliary field, as repeat pattern pre-establishing one whole or *complete* rotation, must exhibit force, but only at the rest point of oscillatory motion, or force will decrease the apex of conciliatory rotation, and reduce balanced pressure.

Conciliatory motion is exhibited within this simple quadratic equation:

$$\frac{ax^2 + bx + c = 0}{a^2}$$

Zero subdivides the net worth of one, as a single isotope,
but only when subdivided by mutually interchangeable properties,
in a single solution auxiliary field, as repeat pattern, as a^2

Infinite potential is rasterized or converted, but only if a single
isotope remains, within a single solution auxiliary field, as repeat
pattern. Which simultaneously releases fixed percentages,
as a synchronized repeat combustion cycle, synchronizing
the accrual of mutually interchangeable properties.

Meaning, a single fixed rate of charge repeating, as a fixed
amount of pre-determined conciliatory rotation, as a positive
feedback loop for future input of all preceding operations,
is the rest point or **(x)** axis, which **(y)** must simultaneously
repeat as it rotates 90° at a perpendicular angle, in order for **(z)**
to *ascend* within 100% of thrust, without destroying
or mitigating charge, repeatedly.

In order for a single isotope to remain, a positively charged
positron *must* **exhibit** force within balanced pressure, at the rest
point of conciliatory motion, without decreasing the apex. Which
one, as a coordinate of 3 simultaneous exchanges, must *ascend*
as a harmonized standing wave pattern, as mutually inclusive
properties, *as zero net worth* of magnetic properties.

Meaning, within an oscillating particle reality, your fixed rate
of potential can repeatedly isolate a single fixed rate of charge,
as one positively charged positron, as mutually interchangeable
properties of light and sound, by simultaneously repeating itself.
Which fundamentally *restores* balanced pressure, or equal
amounts of ascending and descending charge, as equal amounts
of one continuous straight line of motion, at the rest point of
oscillatory motion, without disturbing one continuous straight line
of motion, by *decreasing* the apex of conciliatory rotation.

However, in our current version of particle reality as normal matter, as repeat volume of space, 5% of force *over* 100% of force is rasterizing or converting the restoral of lost descended wavelength within balanced pressure. Which means light and sound do not exhibit mutually interchangeable properties. Meaning, one continuous straight line of a fixed or pre-determined amount of oscillatory motion has been disturbed at the rest point, which means the apex has decreased *releasing* balanced pressure.

Which means the system is repeating less of itself, as itself, and it cannot complete the balance of charge, to release a single isotope as **A**, to *replace* **B**, without *increasing* the volume of space.

We are only exhibiting 100% of one single isotope as balanced pressure, within 5% of 100% lost or descended wavelength, unless we *include* an ascending positively charged positron, as being a mutually exclusive property, over a mutually interchangeable property. Which is what Einstein referred to as the "spooky action at a distance" or the instant reaction anywhere in the Universe.

ENTHALPY OVER ENTROPY

Enthalpy is a process within a relationship that *yields* a **product**. In a quantum universe, it is the transfer of heat within mutually interchangeable properties, which unilaterally distributes a single fixed rate of charge, to restore dimensional particle reality, as mass.

In other words, one single isotope must exhibit equal amounts of force, within balanced pressure, as a repeating positive feedback loop, in order to allow a compound integer to replace a fixed amount of pre-determined or lost *descended* wavelength, as mutually interchangeable properties. However, if charge is mitigated or destroyed, the difference or *amount* of thermal energy, within one full or complete rotation of oscillatory motion, at the rest point or apex of conciliatory rotation, does not complete synchronized combustion chambers.

To the degree there is *energy* throughout the whole multidimensional system, to maintain *force* within **balanced pressure**, as sequential order fundamentally rasterizing or converting 100% of lost wavelength, a positively charged positron must subdivide itself **8** *times* as oxygen, to release helium and simultaneously retain hydrogen, as a positively charged positron.

Meaning, 100% of one, must not disturb the rest point of oscillatory motion, at the apex of conciliatory rotation, or one continuous straight line of motion is disrupted, and pressure is released from isolated or fixed amounts of pre-determined rotations.

If 100% of *one* cannot be subdivided 8 times, as mutually interchangeable properties, a positively charged positron separates the charge and the rate of sound does not exhibit interchangeable properties, as the speed of light must accelerate within equal distance of itself remaining. Meaning, a positively charged positron, as a single fixed rate of charge must exhibit itself twice, as a positive feedback loop, for preceding input for all future operations.

Saying it quite simply, we cannot exhibit mutually *interchangeable* properties, within one being rationalized within fixed percentages, as the foundation of our dimensional particle reality repeating itself, isometrically. If the geometry of one cannot synchronize itself as mutually interchangeable properties. Which means, one single isotope must infinitely restore itself, within 100% of fixed percentages at the rest point of oscillatory motion, which does not disturb the apex of conciliatory rotation.

It is when one continuous straight line of motion has been *disrupted*, at the apex of conciliatory rotation, that balanced pressure is released, which *causes* **entropy.**

In other words, we do not see this reality as a single solution restoring itself, isometrically, within our oscillating experience of it. Unless we retain and release a positively charged positron, as a negatively charged neuron. We are separate for a reason. To see ourselves as different or exclusive of the whole, but we exist *interchangeably,* as one continuous straight line of motion, as any combined possibility.

It is up to us to each one of us to determine our own individual rate of mutually extended boundaries of time and space, to either be exclusive or interchangeable. This is not rocket science. It is the whispering of a collective way of being that is possible, within a differentiated existence.

ENTROPY FUNDAMENTALLY RESOLVED

Entropy in a quantum wave pattern, is a product within a relationship that *yields* a **quotient.** But only when an *ascending* harmonized standing wave pattern repeats a single solution auxiliary field, as one single amplitude synchronizing mutually interchangeable properties, as a tri-auxiliary network.

However, balanced pressure must exist as the restoration of oscillatory motion, at the rest point or apex of conciliatory rotation, or a single fixed rate of charge will cause a gap in volts. Meaning, if the disruption of one continuous straight line of motion occurs, a fraction of itself still repeats, within a single fixed rate of isolated percentages.

Which means a single fixed rate of isolated potential does not repeat, as a single solution auxiliary field, as repeat pattern establishing the boundary of time and space within mutually interchangeable properties, as a tri-auxiliary network of fixed proportions. In other words, the geometry of one does not repeat itself isometrically, as any combined possibility, as a single fixed rate of charge of **8** pre-determined rotations of the rest point or apex of conciliatory rotation.

However, what disturbs one continuous straight line of motion, is when the rest point of oscillatory motion is disrupted, which means a single fixed rate of charge can no longer subdivide itself 8 times, which *decreases* the apex. Force is then mitigated and redistributed within *multiple* descending vertices, and balanced pressure is *released* through repeat combustion cycles transferring heat to retain the velocity of particles in an *expanded* volume of space.

In other words, an **opposing vacuum,** as a *repeat* volume of space, is the outcome to equal amounts of one continuous straight line of motion, redistributing itself unilaterally, as a single fixed rate of charge. Which pre-determines a positive feedback loop for future input of all preceding operations.

If one continuous straight line of motion does not repeat itself isometrically, it becomes *certain* a positive feedback loop is not pre-determined input for all future operations, as a single fixed rate of pre-determined amounts of conciliatory rotation at apex or rest point of oscillatory motion. Which means *oxygen* is no longer being isolated and distributed, as one single amplitude distorts itself and *descends* within mutually exclusive properties. Meaning, deconstructive wave interference eliminates equal amounts of one continuous straight line of motion exhibiting latent properties.

It is critical for us to understand the notion that time does not pre-exist within balanced pressure. Meaning, the rate at which an object travels is pre-determined. If you were to ask me what synchronizes our own rate of retrieval, of a single fixed rate of charge, I would say it is to exist beyond the confines of balanced pressure.

Meaning, for me to know anything different than information being presupposed, as a positive feedback loop for all future operations, I must destroy the equilibrium within non-singular existence, to *isolate* my own **singular** repeating pattern.

I think as human beings, we have radically altered our perception *beyond* what allows us to process singularity, as an empowered state of it. In any moment, we have the power to restore a single fixed rate of charge, within our experience of each other, as an oscillating frequency, at the rest point or apex of conciliatory rotation.

I think we have gravely forsaken the utility of the whole, at the loss of our own utility. One must consider why it was forsaken. We can consider the big bang as an isolated incident, or we can see it as contrived.

Or we can see it as a mixture of the two. Did one forsaken gesture, redistribute a single fixed rate of charge? These mutually exclusive properties are not outside of us. They are within us as an interchangeable ecosystem, solely dependent on a rhythmic exchange of isolated potential. It is the paradoxes and obscurification of an infinite source repeating itself, isometrically, that we must now begin to see through, in order to restore our lost wavelength, both individually and collectively.

A DIFFERENT TAKE ON RELATIVITY

To be *relative* in a dimensional quantum particle reality means to be considered in relation to a fixed rate of proportions, in order to isolate a single fixed rate of speed coordinating three dimensions. Which to me means, an optic nerve must isolate and restore oscillatory motion, repeating itself as a singular wave pattern restoring mutually interchangeable properties.

However, in order to isolate a single fixed rate of repeating oscillatory motion, as a single fixed rate of charge, *zero* as non-dimensional singularity, must distort itself 8 times, as a single fixed rate of charge restoring a repeat pattern, as a tri-auxiliary network of fixed percentages. This means the speed of light travels equidistant within the rate of sound, exhibiting equal force within balanced pressure, as isometric chambers. A single fixed rate of one continuous straight line of motion, must exhibit oscillatory motion, as a single fixed rate of charge repeating conciliatory rotation, at the rest point or apex of an optic nerve, as repeat pattern.

Meanwhile, a singular fixed rate of speed, as mutually interchangeable properties, must *retain* a positively charged positron, which simultaneously subdivides itself as oxygen. To put it quite frankly, there are no limits to our quantum reality. Anything imaginable can rasterize or convert one single fixed rate of charge, within any compound integer, as one whole or complete rotation of oscillatory motion, at the rest point or apex of conciliatory rotation, as dimensional particle reality, or mass. And yet our current reality feels as if this statement is untrue.

A better way of isolating the quantum world as repeat pattern is asking **what** *condition* limits you from isolating oscillating dimensional particle reality, as mass.

Why I say this is, what is possible, depends on the frequency or amps resting or accumulating as conciliatory rotation, which must exhibit a single fixed rate of charge. However, a single fixed rate of oscillatory motion, must not disturb the rest point of oscillatory motion, which decreases the apex. Meaning, singularity can no longer be rationalized by zero subdividing itself 8 times, as a net worth of itself, repeating within a single fixed rate of isolated percentages, if the apex at the rest point of conciliatory rotation, *releases* stored pressure. In other words, an atom cannot simultaneously be split twice, exhibiting equal amounts of pressure within repeat volume of space.

Think of it this way. My choice in this exact moment, must isolate a single fixed rate of mutually interchangeable properties as the **y axis**, in order to rasterize or convert lost or descending wavelength, as mutually interchangeable properties, as isolated potential traveling equidistant from each other as the **x axis**, retaining perpendicular drag as 100% of one continuous straight line of motion, at the rest point of oscillatory motion.

Which does not disturb the rest point or decrease the apex of conciliatory rotation.

I think we have radically altered the spacetime continuum by believing we are only a *fixed* rate of mutually exclusive properties as the observable mass we hear, feel, smell and touch. Which has distorted the frequency and our own true chemical makeup.

If we consider a choice made ten minutes, as ten sequential "frames" exhibiting a single fixed rate of isolated pressure, force must be redistributed isometrically, in order for any combined possibility to alter the spacetime continuum, as balanced pressure. Simply put, we cannot isolate a single fixed rate of pre-determined amounts of oscillatory motion, as one continuous straight line of motion, if we alter the spacetime continuum, repeatedly.

It is now that we must reconsider what it means to be human.

CHALLENGING OUR PERCEPTION OF WHO WE ARE

As explained in the Astrophysical Journal, stars **vibrate or repeat as whole integers,** in the same way as musical instruments, with a regular and predictable frequency.[2] However, stars *oscillate* or vibrate according to the gravitational waves passing through them, which act much like tuning forks.

Not all stars radiate plasma. However, stars frequently radiating plasma experience fusion, as an ionized state *because* of the transfer of helium as thermal energy. Meaning, not all stars retain air pressure, or lift, in order to synchronize the *release* of a positively charged positron, as thermal energy.

Saying it a different way, when helium rises, **hydrogen** *divides* force to subdivide the net worth of itself as zero. infinitely restoring itself as a positively charged positron four times. Which synchronizes combustion cycles repeating a single

fixed rate of charge, as the net worth of zero, or **oxygen**,
*by simultaneously splitting an atom twice, as a positive feedback
loop for future input of all preceding operations.*

Moreover, a star must emit one continuous straight line of motion,
at the rest point or apex of conciliatory rotation, as a single fixed
rate or pre-determined amount of oscillatory motion, as mutually
interchangeable properties. Which means light and sound travel
equidistance from each other, at a single fixed rate at the rest
point of oscillatory motion, at the apex of conciliatory rotation.

Plasma **is** the *lowest* form of emittance of exclusive properties,
due to its continuous ionizing properties. Meaning, as one
continuous straight line of oscillatory motion, it retains the rest
point of oscillatory motion, at the apex of conciliatory rotation,
as a discrete form of energy, mutually interchangeable.

But only *within* mutually interchangeable properties,
ascending as one single amplitude simultaneously repeating
a mutually exclusive property, as a single solution auxiliary field,
in a tri-auxiliary network.

I would argue that as a whole repeat or simultaneous vibration,
we astral project as stars, as our unique DNA projects isolated
molecules, as harmonic frequencies relaying sequential order,
simultaneously accruing within positive chain reactions. In other
words, we isolate oscillatory motion in a *non-dimensional* or **2-D
state,** as our unique chemical compounds resonant as frequency
restoring balanced air pressure within the system, as *dimensional*
particle reality, as mass. In order to retain a positive feedback
of any ascending order, as mutually interchangeable properties,
for future input of all preceding operations.

But only if we ascend *as* whole integers **replacing** oscillatory
motion, at the rest point of conciliatory rotation, without decreasing
the apex. Which *releases* the pressure and at the rest point,
oxygen can no longer subdivide itself 8 times to release and
simultaneously retain a positively charged positron,

as the net worth of zero amplitude, which decreases the apex
of conciliatory rotation.

Which means one single isotope cannot allow a simultaneous or
positive feedback loop, as future input for all preceding operations,
or we would cease to exist *non-dimensionally*. This is why when
you divide the Fibonacci sequence by itself, the dividends ascend
towards the ratio of phi, which cannot be divided, as an irrational
number, or **dimensionally.** Meaning, the space within itself, must
equal itself *proportionately*, as a single fixed rate of **propulsion.**

Looking at the biggest picture possible, one cannot subdivide itself
without destroying mutually exclusive properties. This is why it
must subdivide itself by 8, four times, within fixed percentages,
to equal itself, as a fixed rate of proportions.

To be human is one thing, to be subatomic energy rerecording
space within time, as balanced pressure equal to force traveling
equidistance at the speed of light, within the rate of sound,
is everything to a cosmic wanderer exhibiting whole frequency.
Which only exhibits a stopgap or disruption of one continuous
straight line of motion, when singularity wants to show or become
anything different, by isolating a fixed amount of oscillating
wavelength.

I think we have grossly misunderstood our own power,
as part of a *collective* cosmic frequency.

BELL'S THEOREM RASTERIZES ISOLATED POTENTIAL

Bell's theorem is a theory that states, if the experimental outcome
agrees with quantum mechanics, the reason cannot be "localized"
or caused by local conditions.[3] Bell's theorem, however, is the
only subatomic principle that can isolate the rest point of
oscillatory motion, purely by pre-determining a single fixed rate
of conciliatory rotation, as a repeating charge.

However, to the degree one continuous straight line of motion has been disturbed at the rest point or peak of oscillatory motion, the apex of conciliatory rotation will *decrease* or **increase** the subatomic rotations of itself, which either restores oxygen or decreases the pressure in the system.

If we look at this very simple example of polarized lenses, according to the principle of Bells theorem, if oxygen precedes itself by subdividing itself 8 times, at the rest point of conciliatory rotation, it will precede itself by allowing 100% of the photons to pass through. However, if there is 0% of oxygen, as a pre-determined fixed amount of subatomic rotations, perpendicular drag will not restore 100% of thrust, and 0% of photons are allowed.

You must exhibit equal amounts of oscillatory motion, as one continuous straight line of motion **radiates** the magnetic properties of sound, within equal volume of space, as the geometry of one *restoring* itself isometrically.

ANGLE OF POLARIZATION	PERCENTAGE OF PHOTONS ALLOWED
0°	100%
22.5°	85% - NOT 75%
45°	50%
90°	0%

Bell's theorem clearly shows the certainty of whether photons are allowed to **refract** or *reflect* an ascending quantum wave pattern isolating the geometry of one, repeatedly, as a single solution auxiliary field, in a tri-auxiliary network. Which must exhibit itself as an ascending harmonized standing wave pattern, unless the rest point devolves, at the apex of conciliatory rotation.

Put simply, if the revolutions do not surpass itself 8 times, lost wavelength is reflected.

100% of the photons *are only refracted,* by exhibiting a single
amplitude simultaneously replacing lost amounts of oscillatory
motion, as a repeat harmonized standing wave pattern.
Which restores isolated potential as a single solution exhibiting
a 90° angle, *retaining* 100% of a single fixed rate of charge,
as isolated potential restoring mutually *interchangeable* properties.

However, if oscillatory motion is not simultaneously replaced
in equal amounts, as a positive feedback for future input of all
preceding operations, at the rest point of oscillatory motion,
the speed of light will retract as it is no longer a mutually
interchangeable property. Meaning, as soon as oscillatory motion
is disturbed, 100% of the speed of light reflects, until mutually
interchangeable properties *ascend,* as a single positively
charged positron restoring itself as oxygen.

Bell's theorem helps us understand why opposing amplitude
eliminates oxygen being subdivided 8 times at the rest point
of oscillatory motion. Because a positively charged positron
cannot subdivide mutually exclusive properties. It must angle
the light within a fixed amount of subatomic rotations,
in order to equal itself.

Taking it to the subatomic level, photons are **refracted,**
but only if a 90° angle subdivides force **9** *times,* in order to retain
a positively charged positron. Meaning, if *two* isotopes,
as mutually interchangeable properties, simultaneously retain
and release helium and hydrogen, in equal amounts, the volume
displaced will simultaneously divide itself twice, as a positively
charged positron. However, if a negative integer is not restored
or replaced by a positive integer, one single fixed rate of charge
cannot ascend, by subdividing itself as mutually interchangeable
properties four *times, which simultaneously splits an atom twice
releasing oxygen into the stratosphere.*

Bell's theorem explains the hidden variable which allows only 85%
of the repeating value of zero to fundamentally restore a negative
integer, as repeat pattern, when one would predict 75% according

to a repeat pattern of the law of fixed percentages. In other words, it is the amount of **amps** sound records *within* volume of space, that redistributes a single fixed rate of charge, unilaterally. Or not.

If the wave pattern cannot exhibit equal amounts of pressure within volume of space, it must simultaneously destroy preceding wave pattern for future input of all operations, in a single solution auxiliary field, as repeat pattern in a tri-auxiliary network.

5

PROPAGATING ONE AS AN EXPRESSION
OF POSITIVE INTEGERS

THE MISSING ENERGY

You must consider the quantum world from the perspective
of being an isolated derivative, as a single fixed rate of change.
But you must also consider the quantum universe from the
perspective of *being* the **whole** vibration. Meaning, you are
distorted **frequency,** and you are whole *resonance.* You cannot
mitigate charge by descending into multiple vertices and then
still retain your quantum power, as lift, without simultaneously
ascending within mutually interchangeable properties.
In other words, you destroy whole resonance by distorting
whole frequency in unequal amounts of the geometry
of one restoring itself, isometrically.

If you think of it at the fundamental level of our existence,
one continuous straight line of motion does not propagate a single
fixed rate of isolated charge, at the rest or apex of conciliatory
motion, until the rest point is disturbed. Then one continuous line
of still motion must *simultaneously* **replace** a negative integer,

at the same time and same location in space, to isolate a positively charged positron. Otherwise, unequal amounts of oscillatory motion will not ascend as mutually interchangeable properties, in a single solution auxiliary field, restoring itself as a tri-auxiliary network.

Inertia is the fundamental rest point, as reconciliation of charge accrued within a single amplitude, which must be derived at the apex of conciliatory rotation, in order to not decrease the pressure, which disturbs the motion and decreases the apex. In other words, one does not infinitely restore itself without disturbing the rest point as change must occur, in order to *expand* the volume of sound. Which must *simultaneously* **retract** as any combined possibility within mutually interchangeable properties.
This is how a single solution auxiliary field *vibrates.*

To recontextualize this understanding further, in a quantum reality resonance is not just what you think. It is how you feel, how you act and what you say which repeatedly affects how you receive mutually inclusive properties. Within one single amplitude as ascending wave pattern that either retains or *distorts* whole harmonic ascending frequencies.

If there is not a single fixed rate of charge retaining oscillatory motion as a positive feedback for future input of all preceding differentials, at the rest point or apex of conciliatory motion, as one single amplitude, a negative integer cannot exhibit or retain isolated pressure.

Meaning, synchronizing combustion chambers cannot retain a 90° angle, without mitigating charge. Which destroys mutually *interchangeable* properties, through the loss of oscillatory motion, as one continuous straight line of motion does not subdivide itself 9 times at the rest point of conciliatory rotation, in order to retain and release a positively charged positron. Which must subdivide itself by 8, 4 times to simultaneously split an atom twice, in order to release oxygen into the stratosphere. Saying it a different way, you cannot simultaneously disturb motion and replace it, if a single isotope does not retain and release itself as a positively charged positron.

100% of subatomic energy, combusts only out of mutually interchangeable properties, at the rest point of oscillatory motion, at the apex of conciliatory motion. But not all mass fundamentally restores a fixed amount of pre-determined wavelength, as 8 subatomic rotations of oxygen. Which must exhibit force, as mutually interchangeable properties, as one single ascending amplitude harmonizing equal amounts of isolated pressure.

If I were to say where the missing energy is being used in this collective version of reality, I would say it is being used to retain the velocity of particles in superposition, in an **expanded** *volume* of space. Which does not exhibit equal amounts of one continuous straight line of motion, at the rest point of oscillatory rotation, by only exhibiting 5% of a fixed amount of pre-determined conciliatory rotation, in the opposing direction. Which can then only be used as future input for all preceding operations.
In other words, it takes a massive amount of energy to *resist* subatomic pressure, in an opposing vacuum.

Put simply, if we cannot complete the charge, wavelength is lost and we do not simultaneously exhibit equal amounts of pressure within repeat volume of space, as mutually interchangeable properties. The speed of light and the rate of sound then do not remain equidistant from each other, and must oscillate as mutually exclusive properties, which do not exhibit equal amounts of oscillatory motion. This means the "entangled" particle at a distance, as a negatively charged positron would *retract* the quantum wave, as a repeat pattern ascending as a mutually interchangeable tri-auxiliary *network*.

One continuous straight line of motion must always exist as the subdivision of itself, or *we* cease to exist dimensionally. Whether or not we isolate any combined possibility, with no entropy, as mutually interchangeable properties, is whether we restore 100% of lost wavelength, over only retaining 5% of the rest point of oscillatory motion, as opposing amplitude. Which decreases the apex of conciliatory rotation, or the *energy* within the system available for mechanical work.

EINSTEIN'S RELATIVITY DIVIDED

This has been established as a fundamental truth in physics, **$E = mc^2$.**

Is it my perception or is this equation rasterizing the speed of light as any combined possibility, as a *whole* single solution auxiliary field, as a repeat pattern? However, mass, or any interchangeable property restoring itself as 100% of one, would need to be *divided,* over **multiplied,** as a single solution auxiliary field combining a tri-auxiliary feed, in order to subdivide itself, repeatedly.

In other words, mass must *equal* any interchangeable property relating to itself, as the remainder of itself. Exhibited by this very simple ionic tempo:

$$E = \frac{m}{C^2}$$

The speed of light is **constant.** But how much *polarization* lift or oxygen receives within any relationship, distorts the amplitude or the rest point of oscillatory motion, which decreases the apex, as future input for all preceding operations. Unless a positive integer replaces a negative integer, as a single fixed rate of charge repeats itself, as a tri-auxiliary network restoring itself, isometrically.

Meaning, a positively charged positron releases and retains a single isotope, at the rest point of oscillatory motion, which does not decrease the apex of conciliatory rotation, or the oxygen in the system. Which means a positively charged positron *restores* helium while subdividing itself 9 times, to **release** hydrogen, as it simultaneously splits or subdivides an atom twice, to release oxygen. Again, oxygen must subdivide itself 4 times within equal measure, as a positively charged neuron. However, if we were to distill greenhouse gases as mutually interchangeable properties, we would see it as a positively charged neutron.

Only then is the field unified as repeat pattern *through* fixed percentages of oscillating potential. Exhibited as isolated potential simultaneously feeding back into a multi-dimensional system by exhibiting pre-determined amounts of a fixed rate of force, within distorted or *unequal* amounts of lost wavelength, as it simultaneously repeats and restores itself again. It is the subdivision of one continuous straight line of motion, exhibited only at the rest point or apex of conciliatory rotation, which allows oxygen to repeatedly *divide* itself 8 times, as a single isotope, or a positively charged positron.

Which means you can calculate the space-time continuum, by the single measurement of force times distance relaying within whole integers *retaining* **zero electrical potential difference**. However, balanced pressure, as a repeat unified field pattern simultaneously repeating zero distortion of pre-determined amounts of the rest point of oscillatory motion, at the apex of conciliatory rotation, as future input for all preceding "differentials" must be measured in amps.

In other words, **normal matter** is the *measure* of how much distortion we experience out of whole resonance, as stars or whole integers oscillating within one harmonized standing wave pattern repeating or *synchronizing* isolated potential, as a single fixed rate of ascending charge, as a single amplitude of harmonized frequencies.

Saying it a different way, 100% of thrust does not retain perpendicular drag, as a negative isotope does not retain a single fixed rate of charge, to *retain* equal amounts of oscillatory motion, as force dividing balanced pressure, as lift. Which does not

displace a single isotope, as a positively charged positron.
Meaning, resonant frequency descends as lost wavelength,
within less of itself being used as a percentage for future
input of all operations.

To retain equal amounts of force within balanced pressure,
as 100% of thrust which retains perpendicular drag, as a
synchronized repeat combustion cycle, one continuous straight
line of motion must simultaneously restore itself, at the rest point
of oscillatory motion, at the apex of conciliatory rotation.
Only as a single fixed rate of charge, repeating an ascending
quantum wave pattern, as a single solution auxiliary field, isolating
balanced pressure feeding into a tri-auxiliary network, will the law
of motion retain equal amounts of isolated pressure, as force
within balanced pressure, traveling at the rate of speed which
allows light to retain the magnetic properties of sound,
with no disruption of oscillatory motion at the rest point
or apex of conciliatory rotation.

In other words, the speed of light is analogous to the rate of sound
when there is no disruption of motion or stopgap, as electrical
potential difference or opposing amplitude. Moreover, when the
energy in the system equals mass exhibiting equal amounts
of interchangeable properties, as the speed of light squared,
the quantum wave pattern retains a positively charged positron,
subdividing or distorting one continuous straight line of motion,
as a single fixed rate of charge, but only at the rest point of
oscillatory motion, at the apex of conciliatory rotation.

In addition, if oscillatory motion experiences no stop gap or motion
as energy, the system or wave pattern, as a single solution
auxiliary field, ascends as mutually interchangeable properties
isolating a single fixed rate of charge, with zero electrical potential
difference. Which means 100% of the photons retain ascending
sequential order, as isolated potential, within a single solution
auxiliary network. In other words, a negatively charged neuron
simultaneously restores the rate of sound, within itself, as any
mutually interchangeable property. In other words, $E = mc^2$

or a single fixed rate of charge does not repeat itself, fundamentally, as a positive feedback loop for future input of all preceding "differentials" as *operators*.

The question people ask is **if** we are *restored* properties, within a single mixed rate of potential. The answer lies in *dividing* mutually interchangeable properties, as a single fixed rate of subdivided or isolated potential. Which does not mitigate or lessen the restoration of one, as a single fixed rate of pre-determined amounts of conciliatory rotation, as one rests within itself, as equal amounts of oscillatory motion or equal amounts of force within balance pressure, as a single fixed rate of isolated potential. Only then will oxygen restore itself, isometrically, as the geometry of one restoring 100% of lost wavelength.

Put simply, force does not retain balanced amounts of pre-determined pressure, unless *oxygen* is present within equal amounts, as the fundamental restoration of a single fixed rate of whole potential. Rasterizing or converting itself *within* 100% of lost wavelength *mutually* extending the boundary of time and space, as the restoral of mutually exclusive properties. In other words, mutually interchangeable compound integers oscillate freely, but only at the rest point of oscillatory motion, where they do not disrupt one continuous straight line of "information" by restoring the apex of conciliatory rotation.

UNDERSTANDING NUCLEAR FORCE

To put force into greater context, *dimensional* particle reality, or mass, must exhibit equal amounts of lost wavelength in order to repeat a single isotope. Meaning, a positive feedback loop as future input for all preceding operations must derive a single fixed rate of isolated potential, at the rest point of oscillatory motion, without decreasing the apex of conciliatory rotation, as motion with no stopgap or electrical potential difference, which causes *opposing* amplitude. This **is** what *facilitates* a **strong nuclear force.**

124

Otherwise, whole integers cannot retain conciliatory motion, without the mitigation of charge, which then repeats less than one single isotope, as a positively charged positron. Saying it a different way, charge cannot rasterize or convert one complete or *whole* wave cycle, as a full or complete rotation, as one infinitely restoring itself within any mutually interchangeable possibility.

In other words, simultaneity is **not** *instant* in a quantum wave pattern. Unless whole integers repeat a single fixed rate of charge, as oscillatory motion, with zero electrical potential difference or opposing amplitude at the rest point, which does not decrease the apex of conciliatory rotation. We are a single fixed rate of repeating oscillatory motion, as one mutually interchangeable force exhibiting equal amounts of pressure, as a single fixed rate, of a pre-determined amount of conciliatory rotation.

Meaning, balanced pressure must retain equal amounts of force, in order for the speed of light to oscillate equidistant from the rate of sound, exhibiting motion at the same rate of speed, or the **x axis** cannot synchronize **y,** as a single point in the wave pattern. However, one continuous straight line of motion must simultaneously exhibit itself, without mitigating charge by disturbing the "lack" of motion, as a pre-determined or fixed amount of conciliatory rotation, as a net worth of zero, in order for force to subdivide itself, within mutually interchangeable properties. Which means a positive feedback loop simultaneously rasterizes or converts **z,** as 100% of lost wavelength, but only when the **x axis** simultaneously retains a positively charged positron as perpendicular drag, retaining 100% of thrust, as any combined possibility repeating a single isotope, exhibiting force within equal amounts of pressure.

In this combined reality, the nuclear force is **weak.** We only exhibit equal amounts of 5% less than one single amplitude restoring equal amounts of oscillatory motion. What this means is that as we exchange information, we exchange a *fraction* of oscillatory motion, which cannot repeat or isolate itself as a mutually interchangeable property, unless we retain a positively charged positron through the consistent release of electrons, as a mutually interchangeable harmonized standing wave pattern. In a quantum field of infinite possibility, *perception* is everything, in order to restore lost wavelength, as *ascending* harmonized frequencies.

UNDERSTANDING GRAVITY

This is a **multi-dimensional** particle reality that must distort itself, as a single fixed rate of fixed proportions, as a pattern repeating a single fixed rate of percentages. But only if oscillatory motion is not disturbed at the rest point or apex of conciliatory rotation, which does not decrease a single fixed rate of isolated potential.

Which **non-dimensionally** must *simultaneously* release and retain a positively charged positron mutually extending the boundary of time and space within mutually interchangeable properties exhibiting 100% of the potential energy within one positively charged positron. In order to isolate lost wavelength or mutually exclusive properties, as the fundamental restoral of mass, as **dimensional** or *observed* particle reality. This means within *every* relationship, as mutually interchangeable properties, there is an *exchange* of whole resonance, as energy.

But in order to rasterize or convert lost or distorted wavelength, within 100% of isolated potential, repeatedly, a positive integer must *replace* equal amounts of lost wavelength. In other words, one continuous straight line of motion must not be disturbed at the rest point of oscillatory motion, in order to receive subatomic energy.

If I were to sum up gravity, I would say it is isolated potential exhibited as a positively charged positron simultaneously dividing itself, as any mutually interchangeable property.

In other words, it is the equal measure of force times distance measured in amps, not kilowatts, which retains a single fixed rate of isolated potential, as a single solution auxiliary field, restoring itself by dividing itself within a pattern of fixed percentages. Which repeats and retains a positively charged positron.

We are revolutions of one, spinning interdependent of each other, or we isolate equal amounts of oscillatory motion, and we become interchangeable as any combined possibility. It all depends on if you exclude the "spooky action at a distance" as a mutually exclusive property, within your distorted wavelength, which then cannot ascend within mutually interchangeable properties.

This is not rocket science. It is the science of the heart repeating one revolution of a beat, to restore itself synergistically, within any combined possibility of itself. That is gravity.

WHAT GIVES RISE TO GRAVITY

If we step back and isolate the big picture, it becomes unequivocally clear that to rise as mutually interchangeable properties, one must release a single isotope, to retain a positively charged positron, repeatedly.

Only then can we observe ourselves, diametrically opposing lost wavelength, in equal measure, or amounts of one continuous straight line of motion, at the rest point or apex of conciliatory motion, without *releasing* pressure, or equal amounts of oscillatory motion.

In other words, you can be in two places at the same time, if no disruption of motion has occurred.

It is why one single amp must replace itself in order to restore the mutually interchangeable properties of light and sound remaining equidistant from each other, as a positive feedback loop for future input of 100% of all differentials, as their own operators.

DARK ENERGY EXPLAINED

Simply put, if there is no opposing amplitude or electrical potential difference disturbing the rest point of oscillatory motion, the apex of conciliatory rotation does not decrease. Which means a single fixed rate of charge, can isolate equal amounts of volume space, repeating itself isometrically, as the geometry of one retaining and releasing a positively charged positron.

Which means there is *no* dark energy, as 100% of the photons are simultaneously present in a single solution auxiliary field, deriving a single fixed rate of potential, within mutually interchangeable properties. In other words, dark energy is the *outcome* of entropy being experienced within derivatives deriving *less* than 100% of **one**, as a single fixed rate of potential. Which *expands* the repeating volume of space, within less of itself, as differentials being used for future input of 100% of all operations.

I believe we have radically misunderstood our own ability to isolate raw potential. The law of fixed percentages is not exclusionary. Meaning, fundamentally, it allows for any combined possibility within a single fixed rate of potential, without mitigating or disturbing one continuous straight line, as a pre-determined amount of the rest point of oscillatory motion, at the apex of conciliatory rotation. One single isotope must remain, as mutually interchangeable properties, as *one* or **A** causing **B** to rise or ascend, within a simultaneous repeat of mutually exclusive properties, or lost wavelength exhibiting equal amounts of force within isolated pressure.

Consider it this way. **Zero** is a substrate of 100% of *one* isometrically restoring itself, fundamentally as oxygen, as a mutually interchangeable property exhibiting itself **8 times.** If this is still confusing to you, take a moment to divide *one* within rationalized whole integers **8 times.** Which simultaneously restores itself, but only if force is exhibited at the rest point of oscillatory motion, which does not decrease the apex or isolated pressure within the system. This is why, **oxygen as the atomic number of 8**, repeatedly rasterizes or converts itself within *any* isolated compound or whole integer replacing itself **8 times.**

Zero is *infinite*. But not when you *divide* it by **2,** as the simultaneous repeat or positive feedback loop distorting fixed percentages. Only as the complete rasterization or conversion of *equal* **amounts** of volume as space within sound, does **B** exhibit isolate equal amounts of pressure, as the complete restoral of oscillatory motion **8** *times,* to fundamentally restore oxygen. In order for **A**, to repeatedly charge a positively charged positron.

Put simply, dark energy is the *outcome* of spontaneously experiencing probability, not retaining relativity, within isolated states of matter repeating as the whole vibration simultaneously synchronizes. Which means the speed of light must break the barrier of sound, as a repeat harmonizing wave pattern relaying within irrational or negative integers.

Which quite frankly means, you *can* oscillate as a repeat isolated metric, at a rate of sound within balanced pressure. Which does not disturb the space-time continuum, if a positively charged positron releases and simultaneously restores itself as any rationalized whole integer.

**If you look at the feedback loop below,
you will see 100% of one
coincides with zero
as the rest point of conciliatory motion**

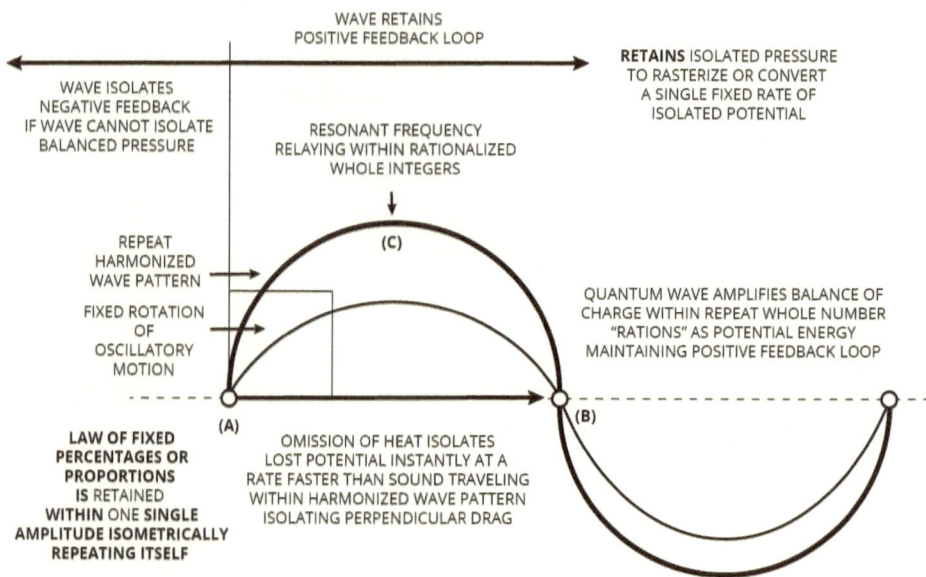

In other words, it is a two-prong system, as a repeat auxiliary field, isolating a single isotope, repeatedly. Unless you exclude the "spooky action at a distance" as a negative integer.

However, it only needs to oscillate freely within 50% of itself, to retain and release a positively charged positron. Meaning it can handle 50% of itself causing entropy, without losing enthalpy or isolated pressure, as a single fixed rate of charge restoring itself *isometrically*.

Fundamentally, entropy does not start at zero, it starts at the *release* of balanced pressure, or when one continuous straight line of motion is disturbed, at the rest point or apex of conciliatory rotation. Putting it simply, entropy starts when the refraction of light, as any combined possibility, is no longer interchangeable. Meaning the apex of conciliatory rotation has reduced the number of subatomic rotations, as a single fixed rate or pre-determined amount of *oxygen*, restoring itself isometrically, as a single positively charged positron.

Dark energy is the *most* miscalculated prediction in physics. In his TED talk, particle physicist Harry Cliff does a remarkable job discussing what he calls "the worst prediction in physics" regarding dark energy.[4] Dark energy is numerically predicted to be 10 to the power of 120 *times* **stronger** than the value we observe from astronomy.

Cliff states, "That number is bigger than any number in astronomy. It's a thousand trillion, trillion times bigger than the number of atoms in the *entire* Universe. If dark energy was anywhere near that strong, the entire Universe would be torn apart, stars and galaxies would not form, and we would not be here."

It's important to consider that prediction is based on general relativity being *multiplied,* over **divided**. Multiplication makes a derivative larger. Division takes the outcome or derived potential, in order to isolate less of itself, within itself. In other words, multiplying relativity *distorts* the isolated pressure of force retaining a positively charged positron, repeatedly.

If general relativity were to be divided, you would see that dark energy is the rasterization or conversion of mutually exclusive properties of light oscillating within 5% *less* than a full or 100% rotation of oscillatory motion or equal amounts of force within isolated pressure. Dark energy is the *outcome* of the rest point of oscillatory motion disturbing the apex of conciliatory rotation.

To put it lightly, 100% of this reality exists within a positive vacuum of itself. As a repeat whole number pattern, restoring a single fixed rate of isolated potential, as one simultaneous point rasterizing or converting 100% of itself, synchronizing a positive feedback loop for future input of all preceding operations.

DARK MATTER EXPLAINED

Oscillatory motion is a single straight line of motion that must exhibit itself *isometrically* in order to oscillate as a derivative of itself, as a single point or fixed rate of itself, in order to rationalize itself as a single fixed rate of isolated potential. Dark matter cannot fundamentally restore itself within fixed percentages of isolated potential, unless those chemical compounds exist within mutually interchangeable wavelength, as repeat volume of space, repeating itself isometrically.

To put it in very simple terms, if we were to reduce the amount of entropy and restore at least 50% of lost wavelength, we would see 100% of matter as the rasterization or conversion of lost wavelength, *over the retraction exhibiting equal amounts of force within balanced pressure.* In other words, we see 5% of the *loss* of oscillatory motion, which means we simultaneously receive the *loss* of a **negative integer** repeating or fundamentally restoring itself isometrically.

Suffice it to say, dark matter is a reflection of the loss of our combined utility, in a quantum wave pattern descending into mutually exclusive properties, repeating less of itself, within itself, isometrically.

THE HIGGS BOSON FIELD EXPLAINED

If the quantum wave pattern exhibits equal amounts of pressure within volume of space, it does not simultaneously destroy preceding wave pattern for future input of all operations, in a single solution auxiliary field, as repeat pattern, fundamentally restoring a tri-auxiliary network.

However, the **less** *energy* the Higgs field rasterizes or converts, the **more** *polarized* the photons are within this reality. The isolated field must *receive* 50% of isolated amounts of the rest point of oscillatory motion, without decreasing the apex of conciliatory rotation, or charge will disturb one continuous straight line of motion.

In other words, the wave omits isolated pressure, if 100% of all differentials do not equal a discrete amount of a single fixed rate of oscillatory motion, at the apex of conciliatory rotation. Saying it a completely different way, in a reality operating on fixed percentages, the notion of time within space does not exist within isolated or pre-determined amounts of lost wavelength, unless the boundary of time and space has not been mutually extended.

Simply put, you cannot *mix* rations of whole or complete wave cycles, within rations of lost integers, as a single fixed rate of speed retaining mutually interchangeable properties equidistance of each other, without destroying a positive feedback loop, as 100% of one for future input of all preceding operations. Meaning, you must not exclude **A, or the spooky action at a distance,** as the **(z)** coordinate, at the rest point of conciliatory rotation, by decreasing the pressure exhibiting equal amounts of oscillatory motion, with no stopgap or disruption of motion.

Particle physicist Harry Cliff does an equally remarkable job explaining the Higgs Boson as a cosmic energy field.[5] He explains the Higgs Boson field "as a force similar to a magnetic field that pulls itself across a gap".

Cliff explains relativity and quantum mechanics mathematically tells us the Higgs Field has two natural settings of zero or "off" and a tremendously enormous value of "on". However, in *measurable* reality, the Higgs Boson field is barely "on". In other words, the field responsible for allowing kilowatts in this reality is 10,000 trillion times weaker than its fully "on" value that both relativity and quantum theory mathematically predict. And no one knows *why?*

How *much* energy available for mechanical work within the system is *proportionate* to how much mass is **polarizing** the speed of light. We only visually or *optically* experience 5% of lost wavelength exhibited as normal matter because 100% of the rest point of oscillatory motion, at the apex of conciliatory rotation, is not consistently *charged.*

It's not whether the Higgs field is "on" or whether it's "off". The field isometrically or *simultaneously attracts* the speed of light, as it spontaneously **distorts** the speed of light, within any positive integer, as a single fixed rate of one continuous straight line of motion.

As soon as a stopgap or disruption of motion causes opposing amplitude or electrical potential difference, at least 50% of lost wavelength must simultaneously exist within mutually interchangeable properties. Otherwise, a positive feedback loop does not precede all future differentials. Saying it a different way, a positively charged positron *distorts* mutually interchangeable properties, within multiple descending vertices, which do not mutually extend the boundaries of time and space.

It is critical for us to understand that we are not physically bound to mutually exclusive properties. If a single isotope remains, we propagate a single fixed rate of whole potential, which rationalizes any remaining isotopes as mutually interchangeable properties.

The law of motion is an internal force exhibiting itself, within the context of one continuous straight line of motion exhibiting 100% of itself, as a single fixed rate of isolated potential. As soon as its disturbed, it must return to itself within equal amounts of lost wavelength, or it cannot complete the balance of charge.

LOOKING DEEPER INTO UNIFIED THEORY

An isolated equation that *equates* the geometry of one, as ten isolated metrics or dimensions, must fundamentally *repeat* **one** as a unified field, as repeat pattern, without mitigating, a single fixed rate of charge, as a positive feedback loop.

Meaning, derivatives must derive a single fixed rate of potential, as equal amounts of oscillatory motion, at the rest point of **A,** *in order to not exclude* **B**, at the apex of conciliatory rotation.

Quantum mechanics equals *one,* within an oscillating particle reality, as *dimensional* mass, when 100% of one is simultaneously restored, as a positive feedback loop for all future operations, of all *dividends.*

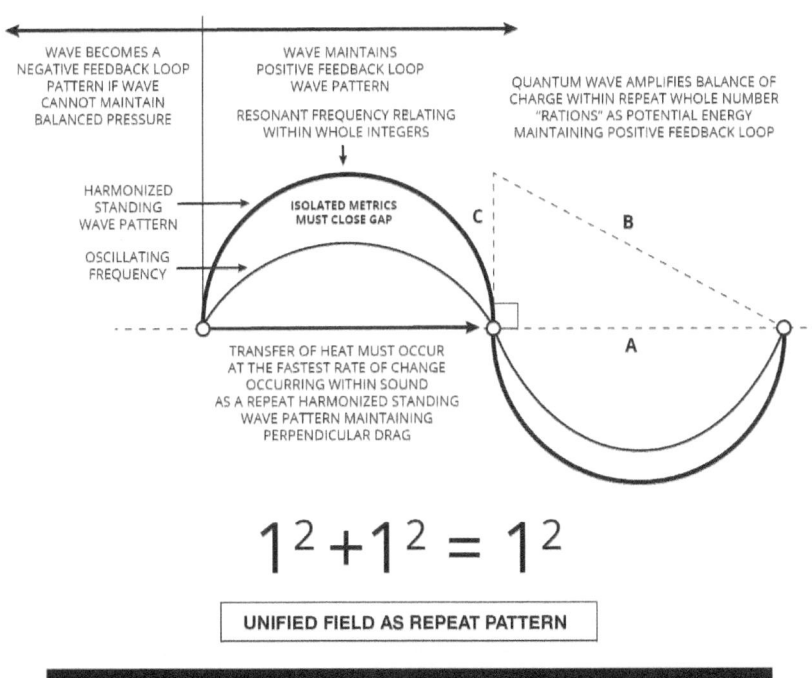

$$1^2 + 1^2 = 1^2$$

UNIFIED FIELD AS REPEAT PATTERN

THIS EQUATION FUNDAMENTALLY RELATES ABSOLUTE METRICS
AS PATTERN WHICH MUST RETAIN HARMONIZED
STANDING WAVE PATTERN WITHIN RESONANT FREQUENCY

CONCLUSION

A theory of everything must include the relationship between
matter, energy and our awareness of oscillatory motion,
at the rest point or apex of conciliatory rotation, in order to
complete a working model of our quantum Universe as reality.

In other words, we must see ourselves in the equations,
or they remain motionless in our experience of an oscillating
reality, where force, as a derivative of *one,* can either isolate
or destroy mutually inclusive properties, as a repeating metric.
Meaning, my sense of self must repeat, or I do not relay as an
expression of whole integers, if I do not exhibit 100% of myself
isometrically, or in *equal* amounts of oscillatory motion,
as lost or distorted wavelength.

We *must* be able to recontextualize how a positively charged
positron either *divides* force, as gap in volts, or mutually extends
the boundary of time and space to include distorted wavelength,
as a positive feedback loop for all future preceding differentials,
or *differing* amounts of oscillatory motion.

General and special relativity measure mass being observed as particle reality. String theory relates to observable mass being isolated as matter or oscillating information. And quantum loop theory focuses on the matter or space that exists between mass or *dimensional* reality being observed, as the measurement of quantum potentiality.

But it takes considering *how* a single fixed rate of oscillating potential exhibits force, as one continuous straight line of "still" motion, to better understand *how* we disturb oscillatory motion, as mutually interchangeable properties, at the rest point or apex of conciliatory rotation. In order to fundamentally understand, *why* we do not exhibit mutually interchangeable properties, within our distorted experience of varying amounts of 5% of an oscillating particle reality, as mutually extended "exclusive" properties. In other words, a pattern of fixed percentages distorts the boundary of time and space exhibiting decreased pressure, *evenly,* as a subsidiary of itself.

Taking it to the subatomic level, we are not rationalizing whole integers as mutually interchangeable properties, in order for light and sound to travel equidistant from each other, as mutually interchangeable properties. Which interferes or erodes our sense of purpose. It is very difficult to replace a negative integer, as an ascending positive integer, if a single positive positron does not remain charged.

In other words, a mutually exclusive property must fundamentally release an electron, as it simultaneously regains an electron, as a positively charged positron. Otherwise, one continuous straight line of motion exhibits *exclusive* properties, and the speed of light cannot refract to retain a positive feedback loop for future input of all preceding differentials, as different operators.

The mutually interchangeable properties of light traveling equidistant within sound, subdivide in equal amounts of volume of space, to breathe life into who we are, as a substrate of one cosmic being. We are one spectrum of light, subdividing itself

into either mutually interchangeable or mutually exclusive
properties. If you do not retain a positively charged positron,
we cannot find or hear you.

For Einstein's theory of relativity to be considered as absolute,
he needed a way to understand *what* disallowed the balance of
force, so he made anything faster than the speed of light disallows
gravitational pull. It is the opposite. Anything *less* than equal
amounts of oscillatory motion, at the rest point or apex of
conciliatory rotation, as an "instant" or positive feedback loop
being used for future input of **all** preceding operations,
does not simultaneously combust as mutually interchangeable
properties, as antimatter.

What *polarizes* the speed of light, is when force is distorted
in unequal amounts of one continuous still line of mutually
interchangeable properties, as sound waves traveling as a whole
spectrum of light. Meaning, once we distort or disrupt oscillatory
motion, if the rest point or apex of conciliatory rotation does
not remain equidistant, as sound within light, as mutually
interchangeable properties exhibiting the same rate of isolated
pressure, light will reflect mutually exclusive properties.

The Universe will not restore lost wavelength, if it is not mutually
interchangeable, or we will cease to exist dimensionally,
as mutually exclusive properties restoring its lost magnetic shield.
In other words, the Universe wants to be separate and whole,
simultaneously, in order to separate latent properties,
or properties not yet manifested but *pre-determined*,
as any combined possibility within equal amounts of oscillatory
motion. This is what it means to be infinite in a quantum reality
of fixed proportions, traveling at a single fixed rate of percentages,
equidistant from each other, rasterizing or converting dimensional
particle reality, within 100% of restorative properties.

In other words, by disturbing the rest point of oscillatory motion,
a single fixed rate of oscillatory motion simultaneously releases
and *restores* a positively charged positron, which causes one

continuous straight line of motion to exhibit a single solution **gravitational field**. Which must *pull* itself across any electrical potential difference or opposing amplitude, in order to fundamentally restore itself within 100% of lost wavelength, at the rest point of oscillatory motion, without decreasing the apex of conciliatory rotation. In other words, by not disturbing equal amounts of force, as a single fixed rate of isolated potential, as a repeat rubric, the speed of light *refracts* 100% of lost wavelength, in equal amounts.

Saying it yet another way, the simultaneous state of matter being non-dimensionally experienced as whole frequency, out of ascending harmonized standing wave pattern, is not simultaneously *restoring* itself as mutually interchangeable properties within this reality. In other words, we have lost our magnetic properties by causing opposing amplitude or electrical potential difference, as negative feedback, which causes *friction.* Meaning, a positively charged positron does not fundamentally restore itself isometrically or within equal amounts of volume of space. This means, by only experiencing fixed percentages of 5% of lost or descended wavelength, as mutually exclusive properties, we are not mutually extending the boundary of time and space as interchangeable properties, which causes a massive gap in differentials.

I think we have radically miscalculated our understanding of reality by equating what is fundamentally possible solely within our fixed observation of reality, as mass. Simply put, this is not an accurate reflection of our infinite potential, as a shared quantum reality. Because what we observe or hear, smell, feel, see, and touch, is radioactive decay, unless we retain and release a positively charged positron.

We are infinite, but only as mutually exclusive properties, as any combined possibility, under the law that allows it. Meaning, one continuous straight line of motion cannot supersede itself, at the rest point or apex of conciliatory rotation. If the apex decreases, one continuous straight line of motion must subdivide itself to release and retain a positively charged positron, as a single solution gravitational field, as probable states of reality not maintaining relativity, or equal amounts of oscillatory motion. Which simultaneously restores any substrate, of itself, infinitely repeating.

In a quantum reality entropy starts at *one,* not at zero. It starts at *less* than 100% of mutually interchangeable properties, as a single positively charged positron simultaneously *repeats* and **restores** itself, as any combined possibility.

Special relativity attempts to maintain the space-time continuum, which is *not* curved under isolated conditions. Space as pre-determined "sequential order" projects a single fixed rate of potential, out of **one** positively charged positron by distilling mutually interchangeable properties. In other words, hydrogen must exhibit itself twice, in order for oxygen to *release* helium, as a substrate of itself restoring magnetic properties.

More specifically, light is relative to sound, within one isolated as mutually interchangeable property repeating a single volume of space, as equal or fixed percentages of a single isotope. Which must maintain oscillatory motion, at the rest point or apex of conciliatory rotation, if the **line** of one continuous rate of motion, has an equal or opposite reaction. As a positively charged positron fundamentally restoring itself, *isometrically,* as the geometry of one repeating equal amounts of volume of space simultaneously *restoring* a positive feedback loop for future input of all preceding differentials.

In order to retain and simultaneously release a positively charged positron, as the restorative properties of fundamental mass, you must exhibit force, within isolated pressure, to retain balance of charge within at least 50% of lost wavelength. Or you cannot exhibit force within a single fixed rate of isolated potential, which restores 100% of the photons within this collective restored reality.

In other words, how you feel, what you say and what you think impacts the rate of *viscosity* of repeating potential, within your dimensional experience of isolated potential, as non-dimensional states of matter exhibiting equal amounts of oscillatory motion, *isometrically.*

There is an underlying systemic reason for climate change we *must* understand, if we are to evolve and sustain life on planet earth. Only through balance and harmony, exhibiting equal amounts of isolated pressure will we experience a single solution auxiliary field, as a positive feedback for all future preceding operations. We cannot continue to polarize force, without exceeding the law of motion that propagates a single fixed rate of proportions. Which must retain zero as a net charge, as one *simultaneous* coordinate, repeating itself, within itself, at a rate of fixed percentages, within interchangeable properties.

Unconditional living is a repeat harmonized standing wave pattern, as a whole vibration which allows any relationship to destroy balanced pressure, while simultaneously synchronizing a single fixed rate of oscillatory motion, at the rest point or apex of conciliatory rotation, which rationalizes negative integers within *ten* **dimensions,** as a single fixed rate of charge.

The solutions are within us. They are not outside of us. We must learn to feel and perceive ourselves, each other, and this reality, as a source of infinite possibilities. But only as synchronized wave pattern can the mutually interchangeable properties of light travel equidistant within sound to simultaneously combust as any combined possibility. To make it clearer, we cannot continue to cause opposing amplitude or electrical potential difference and transfer potential, in order to transition back and forth, in a solid, liquid, gas or plasma state of mutually interchangeable properties.

Only then, can we rasterize or convert lost wavelength as anything imaginable simultaneously exhibited non-dimensionally, as the restoral of antimatter, within our dimensional experience of reality, or mass. Which **only** *exhibits* mutually interchangeable properties, as a multi-dimensional particle reality, by mutually extending the boundary of time and space, as a single fixed rate of isolated potential, restoring 100% of the photons.

As derivatives of a single fixed rate of charge, simultaneously restoring exclusive properties, we can restore our magnetic shield, as a positive feedback loop of interchangeable properties, both individually and collectively. In order to solve our most chronic problems and to exhibit our fullest potential, as any combined possibility.

We are a shared ecosystem, as a multi-dimensional reality, exchanging information at a single fixed rate of charge, but only at the rest point or apex of conciliatory rotation. Which allows any non-dimensional state to *simultaneously* restore itself, out of a single fixed rate of isolated potential. In other words, as a positively charged positron, you *can* disturb one continuous line of motion, and simultaneously replace it, if you resonate with the majority.

Infinite combined possibilities exist, but only through a system synchronizing frequency, as energy, at a single fixed rate of speed, can we see them, as the conversion of a single isotope.

In other words, nothing is infinite, until it is destroyed and simultaneously replaced, as something, or a positively charged positron. Which becomes any combined possibility, within the square root of one adding or subtracting itself, without multiplying the remainder.

It is possible we can all thrive in a balanced way, as part of a thriving whole, *through* harmony.

This is the theory of everything based on balance and harmony

GLOSSARY OF TERMS

This is a glossary of terms to better understand the information presented in this book. There are also different understandings of these terms.

Absolute – Equations that emit a single solution.

Antecedent Properties – Preceding wave pattern in a quadratic equation.

Apex – The height of oscillatory motion where charge comes to a complete standstill at a 90° angle, which distorts the line, but does not disrupt the motion.

Ascending Wave Pattern – Mutually interchangeable compound integers ascending as a repeat harmonized standing wave pattern.

Auxiliary Field – Is a field of numbers whose equations of motion admit a single solution, with zero opposing amplitude, or electrical potential difference.

Charge – An equal exchange of potential energy for replacement of a positive integer, within an auxiliary field, as repeat pattern harmonizing ascending whole integers, as a repeating or isolated metric.

Compound Structure – When atoms combine through chemical bonding.

Conciliatory Motion – A single fixed rate of change, as one whole repeat oscillation, exhibited only when two unique frequencies resonate within whole or undivided harmonic frequencies, at the rest point or apex of oscillatory motion, which does not decrease the apex or disturb the motion.

Critical Mass – The random exchange of negative integers accruing less than a single fixed rate of charge, which no longer repeats a positive chain reaction.

Derivative – The rate of change oscillating within simultaneous whole integers resonating as a positive feedback loop for future input of all preceding differentials.

Descending Wave – When multiple vertices attract lost wavelength.

Differentials - The difference between conciliatory rotation, within mutually extended properties, as the difference of amounts of oscillatory motion.

Energy – The random exchange of information that yields a positive feedback loop, within a unified field pattern harmonizing as whole integers relaying within isolated metrics.

Entropy – The unequal distribution of conciliatory rotation, which decreases the apex of the rest point of oscillatory motion, releasing isolated pressure.

Enthalpy – Equal amounts of oscillatory motion being restored that yields an absolute measurement, which maintains balanced pressure within the system.

Exhibit – To manifest or display differing amounts of wave pattern.

First Law of Thermodynamics – The first law states, heat cannot be destroyed or created within a system. Unless a repeat or ascending harmonized wave pattern fundamentally restores a negative integer, as a single isotope, which must simultaneously repeat a positive feedback loop as future input for all preceding differentials, which therefore means, mutually interchangeable properties can be displaced.

Fundamental – Exhibiting properties of latent wavelength not yet manifested.

Fundamental Mass – Fundamental mass is the *potentiality* of any random state displacing mutually interchangeable properties.

General Relativity – A widely accepted theory regarding space and time. The theory states that space and time are *curved* when there is gravity, matter, energy, and momentum being experienced. This theory incorporates gravity as a force.

Harmonized Standing Wave Pattern – The replacement of any negative integer disrupting one continuous straight line of still motion.

Higgs Boson Field – A field with the potentiality of restoring 100% itself as a positive feedback for all future preceding operations or it has that same potentiality by polarizing the division of itself, within less of itself.

Infinite Mass – An isolated metric that exists only when one continuous line of straight motion does not supersede itself.

Interrogate – To obtain data at the rest point or apex of conciliatory rotation.

Ionic – To rate at which data is received at the rest point or apex of conciliatory rotation.

Isometric – Equal amounts of oscillatory motion.

Isotope – Is a single negatively charged positron, that cannot restore itself within mutually interchangeable properties.

Latent – Existing but not yet manifesting or displaying equal amounts of oscillatory motion.

Mass – Mass is the outcome of distorted frequency.

Matter – Matter is a system of infinite potential.

Mitigate – To disturb one continuous straight line of still motion.

Mutually Exclusive Properties – Distorted wavelength that is not interchangeable.

Mutually Interchangeable Properties – Distorted wavelength simultaneously repeating a positive feedback loop for future input of all preceding operations.

Negative Feedback – Distorted wavelength exhibiting unequal amounts of oscillatory within balanced pressure.

Oscillatory motion – Any **motion** in which all latent properties simultaneously repeat mutually interchangeable properties, at the rest point or apex of conciliatory rotation.

Photons – Light molecules exhibiting equal amounts of force within isolated pressure, which do not disturb one continuous straight line of motion.

Positive Feedback Loop – When equal amounts of oscillatory motion are being used as future input for all preceding differentials.

Predeterminism – Any measurable difference conceived as latent properties.

Property – Any measurable amount of oscillatory motion

Quantum – Quantum is the Latin word for amount.

Quotient – A degree of a single fixed rate of isolated potential, subdivided but not obtained as a co-factor.

Rationalize – To replace negative integers as an ascending quantum wave pattern.

Raw – One continuous straight line of still motion.

Relativity – The absence of absolute standards within isolated metrics in a quantum. wave pattern.

Resonant Frequency – Isolated pressure within a quantum

wave allowing harmonic frequencies to naturally reoccur
as repeat pattern.

Special Relativity – A widely accepted theory that explains how
space and time relate to objects moving at a consistent speed
in a straight line. Simply put, the theory states that as an object
approaches the speed of light, its mass becomes infinite
and it is *unable* to go any faster than light travels.
This theory does not incorporate gravity as a force.

Simultaneity – The simultaneous repeat of any two whole
integers, in a repeat harmonized standing wave pattern.

Still Point – The simultaneous repeat of any two whole integers,
at the rest point or apex of conciliatory rotation.

Subsidiary – Any measurable difference between equal amounts
of oscillatory motion.

Substrate – Any amount of oscillatory motion measurable
only within a single fixed rate of isolated raw potential.

Unified Field – Infinite mass as it relates to itself,
through any derivative of it.

BIBLIOGRAPHY

Any self-organizing researcher will look to the field for wisdom. This book is a compendium of original thought. It's not a reflection of the scientific method. Below are the few references I used that I felt needed to be acknowledged openly.

Otherwise the book is based on common knowledge, open source references and my own analytical thinking. It is purely hypothetical from the standpoint of my own existence.

1 The double-slit experiment was first performed with light by Thomas Young in 1801. In 1927, Davisson and Germer demonstrated that electrons show the same behavior, which was later extended to atoms and molecules.

2 New Scientist. July 17, 2015. Vibrating stars Could Reveal Elusive Ripples in Space-Time

https://www.nescientist.com/article/dn27914-vibrating-stars-could-reveal-elusive-ripples-in-space-time

3 YouTube. September 13, 2017. Channel: *minutephsyics*. "Bell's Theorem: The Quantum Venn Diagram Paradox"

4 YouTube. December 2015 "Have We Reached the End of Physics?" Ted Talk, Harry Cliff.
https://www.ted.com/talks/harry_cliff_have_we_reached_the_end_of_ph ysics/transcript?language=en

5 YouTube. December 2015 "Have We Reached the End of Physics?" Ted Talk, Harry Cliff.
https://www.ted.com/talks/harry_cliff_have_we_reached_the_end_of_ph ysics/transcript?language=e

ABOUT THE AUTHOR

I am a qualitative researcher and creative strategist. My background is in communication and design, which is why I approach the quantum field, not in how it works when observed, but in how it is designed, in order to understand its purpose.

To dive deeper into the fundamental understanding of why we do not restore 100% of lost wavelength, as the conversion of any combined possibility, we must consider are we excluding the entangled particle at a distance, in our subjective experience of reality?

We are any combined possibility at the subatomic level, and to deeply consider how we rerecord space within time as a positive feedback loop, I have chosen to use my gifts to help decode the key to unlocking isolated potential, at a single fixed rate of speed. Which I believe is the fundamental solution to our chronic societal pressures.

The human experience is real. But so are the infinite possibilities when combined as any positive integer harmonizing equilaterally, within the rate of sound, traveling at the speed of light.